CARROT CITY

CARROT CITY

CREATING PLACES FOR URBAN AGRICULTURE

Mark Gorgolewski, June Komisar, and Joe Nasr

THE MONACELLI PRESS

Library of Congress Cataloging-in-Publication Data

Gorgolewski, Mark.
Carrot city : creating places for urban agriculture / Mark Gorgolewski, June Komisar, and Joe Nasr. — 1st ed.
p. cm.
Includes bibliographical references.
ISBN 978-1-58093-311-7 (hardcover)
1. Urban agriculture. I. Komisar, June. II. Nasr, Joe. III. Title.
S494.5.U72G67 2011
635.091732—dc22
                          2010050749

Printed in Singapore

10987654321
First edition

www.monacellipress.com

Design by Think Studio, NYC

CONTENTS

FOREWORD

This book provides a timely overview of practical and theoretical projects that support urban agriculture. As cities across the world seek further evidence for the impact of urban agriculture and policy guidance in response to it, *Carrot City* adds new knowledge to an ongoing revisioning of the contemporary city.

Driven by global imperatives such as climate change mitigation, more equitable economic models, and dietary health concerns, urban agriculture has in the past few years moved from an issue at the edge of public discourse to one at its center. While a long-established literature documents and advocates for urban agriculture in developing countries, the rapid shift of interest in urban agriculture that has taken place in North America, Europe, and Australasia is truly remarkable.

Consequently we are now talking about urban agriculture in new terms, as something strategic and infrastructural: the question today is how a significant amount of urban agriculture can be reintegrated into cities. Reintegration is an important term here; cities have included productive spaces before, and the economic and agricultural logic for locating fruits and vegetables close to a city's center was established by Johann Heinrich von Thünen as long ago as 1826. While historic models should not be romanticized and, by some accounts, did not create particularly pleasant places, they do present examples of closed-loop, zero-waste, and energy-efficient systems. Our task now is to rethink and redesign better ways of doing this.

Carrot City offers a well-balanced overview of such "better ways." Its case studies—built as well as conceptual—celebrate the diverse story that now informs our collective understanding of a productive urban landscape. The drivers for urban agriculture initiatives are very complex; they are also additive, together building a compelling case in its support. From a personal perspective, *Carrot City* extends the Continuous Productive Urban Landscape (CPUL) concept that we articulated in 2005, where we advocated urban agriculture as an essential element of sustainable infrastructure, which can be designed to form part of a coherent urban landscape infrastructure. In *Carrot City*, we can now see the emergence of the components that will form part of the infrastructure for a CPUL City.

Interestingly, architects have led the way in developing propositions for urban agriculture. For people like Joe Nasr who have been thinking about urban agriculture for a long time, the situation is very different even compared to just five years ago, as evidenced, for example, by the London Assembly's report *Cultivating the Capital* (2010) or the German Ministry of Building's search for strategies for "re-naturing cities" (2008). Mark Gorgolewski's work with the Canada Green Building Council demonstrates that a new generation of engineers and designers are searching out this knowledge. Whereas historic models evolved out of necessity, in the contemporary city we still have a window of opportunity to plan coherent strategies for the introduction of urban agriculture.

Carrot City presents us with prototypes for the components that can, together, contribute to the environmentally productive city and will, in time, become further refined and developed. June Komisar has compared the introduction of urban agriculture into cities to the introduction of public sanitation, so that in the future, it may be as unusual to find a city without productive urban landscapes as it would be today to find a house without plumbing.

Carrot City fits well into a growing body of work addressing urban agriculture, from Smit and Nasr's seminal *Urban Agriculture: Food, Jobs and Sustainable Cities* (1996) to emerging work related to design theory. This book—by architects and of relevance to planners, designers, policy makers, activists, and entrepreneurs—will be a useful and significant tool as we systematically examine how cities of the future can become more sustainable.

André Viljoen and Katrin Bohn
Bohn & Viljoen Architects
Berlin and London, 2010

INTRODUCTION

For decades agriculture has been largely absent from Western cities. Urbanites often ignore the place that food production can occupy within the boundaries of densely populated areas large and small, yet metropolitan areas include vast suburban or exurban spaces in addition to unused rooftops and parcels of waste land where significant agricultural activity could occur. Food instead arrives to the city from hundreds or thousands of miles away with its nutritional value, freshness, and flavor having been diminished by pesticides, transport, and processing. Urban spaces have a huge potential to contribute to their inhabitants' food supply, and there is growing interest in exploring this potential.[1]

Architects and designers have only recently considered the promise of urban agriculture[2] as a driver for developing new types of urban places and buildings, along with their capacity for contributing to locally based food systems. The emerging and often grassroots alternative-food movement has also barely engaged with the possible contributions the design and planning professions are capable of making to the reintroduction of food systems to urban space. The physical and spatial aspects of food production, processing, distribution, and marketing are where food issues interface with urban planning and design. Opportunities exist for creative cooperation between planning and design professionals and those who focus on urban agriculture and food systems to reconcile constraints that range from zoning restrictions to construction codes.

Professionals of the built environment, along with invested lay parties, can advocate for access to food sources for the purposes of both nourishment and community-building by working to incorporate farmers' markets, greenhouses, edible landscapes, and community gardens into design and neighborhood planning programs. Connections between food issues and built forms have the potential to transform the components of the food system as we currently recognize it, along with basic assumptions about the nature of programming required in plans for urbanized areas and the designs for many building types, such as schools, single and multifamily housing, and other places where food consumption occurs.

Professional bodies have also collectively begun to recognize food systems as an important area of action recently, as highlighted by the American Planning Association's adoption of a policy guide on the topic.[3] Architects as a group have been slow to bestow importance on food production spaces in designs, though the topic gains more attention, and a population of architecture students who have been attuned to environmental initiatives from a young age have begun to enter the workforce and contribute their ideas as professionals. The increasing emphasis on sustainable design and planning, through programs such as LEED (Leadership in Energy and Environmental Design),[4] has helped to encourage recent projects' design to include innovative energy approaches, water-saving irrigation, green roofs, living walls, and other elements that are compatible with policies for more sustainable food and agriculture systems. This in turn has spread awareness about environmental initiatives and food systems to the public as they engage with these spaces.

AIM AND SCOPE

This book was formed out of the growing consciousness about food production and the potential for this to change our buildings and cities. Carrot City as an initiative in its own right has its roots at Ryerson University's Department of Architectural Science in Toronto. Informal advising to architecture students as early as 2006 about food-related projects led to a symposium in May 2008,[5] a traveling exhibit first shown in Toronto in early 2009,[6] a comprehensive Web site,[7] and now this volume. It includes selected projects that are recently completed or currently underway and visionary, speculative proposals by professionals in architecture, industrial design, sculpture, landscape architecture, urban design, urban planning, enthusiasts, and students. These projects are presented in five sections, representing in part decreasing scales of analysis: Imagining the Productive City, Building Community and

Knowledge, Redesigning the Home, Producing on the Roof, and Components for Growing.

The content presented explores how the design of cities, buildings, and urban gardens can facilitate the production of food in populous areas. It considers the contributions that design professionals and local food advocates can make to strengthen the links between urban environments and food supply, and the impact that agricultural issues have on the design of urban spaces and buildings. Through the presentation of an extensive range of conceptual and practical case studies, it seeks to address the following questions, among others:
- What is the place of food production in the city?
- How can design reflect emerging movements that encourage us to consider ourselves coproducers, rather than consumers, and engage us in the food production and supply process?
- What will increasingly dense cities and built forms look like when we start to habitually design them around a sustainable food infrastructure?
- How are underutilized urban spaces being transformed by current food production projects, and what further untapped opportunities for food production exist?
- What are the implications of urban agriculture for building materials, technologies, and structures?

This book presents North American projects heavily, with a particular emphasis on Toronto and other Canadian cities as well as New York City. While this can be explained partly by the history of the Carrot City initiative itself, it is also a reflection of the thriving urban agriculture movement in these cities, which has resulted in increasingly significant projects that have already made a visible impact on their built environments. The book also includes additional relevant examples from around the world. The authors acknowledge that the geographic scope is limited and that, with numerous exciting examples of creating places for urban agriculture emerging every day on all continents—particularly in developing countries where urban agriculture is practiced in rapidly expanding cities out of sheer necessity—many important projects could not be included here.

THE HISTORY OF PLANNED PRODUCTIVE CITIES AND BUILDINGS

Food availability has been a driving force in the creation of human settlements. Originally, food sources were closely tied to urban forms, since most of them were local or regional. With the rise of industrialization and agribusiness, cheap transport, and food preservation technology, however, the distance between farm and market has increased dramatically and steadily, lead-ing to a substantial attrition of these ties in Western cities. The forms and patterns of built settlements no longer reflect their food supply systems, and spaces for food production within urbanized zones have all but vanished.

Questions of food supply were accorded great importance by early theorists of modern urbanism. This is brought out most vividly in the case of Ebenezer Howard's Garden City theory, which was paramount in the modern planning movement. In each of Howard's cities, five-sixths of the land was dedicated to food production. Howard expected that the generous—for the

United Kingdom—residential plots of 20 by 130 feet would be sufficient to feed a typical family. Howard's ideas, as crystallized in his 1902 treatise *Garden Cities of To-Morrow* and illustrated in his generic planning diagrams, are anchored in creating intricate and complementary relationships between agriculture and the city.[8]

Such relationships are also central to the theories of Patrick Geddes, another influential urban theorist of the early twentieth century. The centrality of agriculture was still of intense interest to urban planners in the years between the two world wars, as seen for instance in the proposals of the Regional Planning Association of America (RPAA). These fleshed out Geddes's con-cept of introducing a transect—a section through a region that describes the land use along its length—to establish specificities of productive uses connected to location relative to city centers.[9]

While reflecting a fundamentally different vision, Frank Lloyd Wright's Broadacre City, as spelled out in *The Living City*, called for agriculture to be integrated into dispersed low-density living. Here families would be settled on one acre of land—his estimate of how much is needed for self-sufficiency—and would be connected to the decentralized city by car and flying machines.[10] Similarly Le Corbusier, in his 1922 Contemporary City proposal, included three types of food-producing areas: "protected zones," where he envisaged large-scale agriculture; large kitchen gardens for detached suburban homes; and 10-acre groups of allotments cultivated by apartment dwellers, with supporting buildings, such as storehouses. Le Corbusier also addressed management of the largest plots and proposed that one farmer would be in charge of overseeing the intensive cultivation for every hundred plots, which would feed the population for the greater part of the year.[11]

During World War II, urban spaces in many countries were transformed into remarkably productive areas, and, more broadly, it was suddenly recognized that food production relied on systems that required planning to achieve predictable and steady supplies.[12] Historical examples from the *potager,* or kitchen garden, to the victory gardens and current examples of cities such as Havana, Cuba, where significant amounts of food are produced, prove that urban food production is a viable and sustainable alternative to shipping food from a distance. This concept of food production as integral to the functioning of cities quickly evaporated after World War II, however, as agribusi-ness and commercial food systems took over production from individuals. Consequently professionals who designed, planned, and managed Western cities ceased to include spaces for food production in their designs and master plans. Agricultural policy became oriented to global markets, and agriculture became fully disconnected from urban policy and urban design. At best, zoning commissions ignored it—ordinances and bylaws of metropolitan districts slated areas formerly designated as agricultural as land reserves for future development, and they literally appeared as colorless voids on zoning maps. At worst, urban zoning hindered food production and sought to exclude it from the ideal of the clean, hygienic, and modern city on the grounds that it was unsightly or unhealthy. This attitude is exem-plified by the haphazard banning of livestock from urbanized areas and bylaws restricting what can be grown on front yards, many of which are still in effect today.

Food production is still struggling to be recognized as a valid use of land and an integral urban function in its own right. Even recent proposals, such as the New Urbanist SmartCode, are still failing to reflect a genuine recognition of green spaces as potentially productive spaces.[13] Nevertheless urban agriculture is slowly reclaiming its place in the city.

THE FUTURE OF PRODUCTIVE CITIES AND BUILDINGS

The separation of cities from their food sources is directly linked to many of the most pressing problems in the world today—including climate change, obesity, pollution, security of the energy supply, and global poverty. Concern about current industrial food systems relates them to damage caused to the natural environment, low nutritional quality, and the high energy consumption necessary for transporting food long distances. Furthermore the question of how to feed an urban population, particularly during crisis, must be confronted. These issues are becoming more urgent every day as neighborhoods devolve into food deserts—locations where affordable, nutritious food is difficult to obtain.

Even in the richest countries, food security is an issue.[14] According to the United States Department of Agriculture, a total of 50 million Americans were food-insecure in 2009, including more than 17 million children.[15] At the same time current concerns in much of the world are about the obesity epidemic, the ballooning worldwide health crisis caused by excessive weight gain due mainly to a poor, high-fat diet. The expansion of urban agriculture is often linked with other social equity trends that seek to reverse the concentration of agricultural production into fewer and fewer hands. This means that government and other agencies are beginning to support small-scale, community-based initiatives that empower urban dwellers to take control of their food supply and teach them about healthy, nourishing, culturally appropriate, and locally grown food.

The reduction in the availability of cheap energy supplies, in particular oil, will have a significant impact on food supply, given the energy intensity of current food production methods and the use of fertilizers based on fossil fuels. The World Wildlife Fund estimates, for example, that 30 percent of the total greenhouse gas emissions in the United Kingdom are a result of commercial food supply chains.[16] With a current world population of about 7 billion and predicted to grow to 9 billion by 2050, modifying what we eat, how we produce it, and where it comes from is a necessity to combat climate change. By reducing the distance between producer and consumer, urban agriculture can lessen energy use. Moreover cooking from basic, fresh ingredients is less carbon-intensive than cooking with highly processed foods. In the United Kingdom, it has been suggested that carbon dioxide emissions could be reduced by about 22 percent if food were produced organically, consumed locally, and grown only to be eaten in season.[17]

In a world where food supplies are increasingly difficult to guarantee, food is expensive to produce, and shipping methods are detrimental to the environment, locally based food systems will be essential to cities' resilience, their ability to respond to crises, and to their effective operation. Growing food in cities reduces the dependency on distant food supply chains that can be disrupted for any number of reasons. Many feel that the significant rise in food prices during 2008 and 2011—due to a variety of world economic and climatic factors—foreshadows future food supply problems. Apart from enhancing food security and reducing our ecological footprint, urban agriculture can also play a role in greening cities, which benefits water management, air quality, and social programs.

Movements such as community-supported agriculture, farmers' markets, the 100-Mile Diet, and Slow Food put the local food supply at the heart of urban sustainability and seek to reconnect cities to their food systems. Local food production and related actions—processing, selling, cooking, recycling—can also act as a focus for community participation and engagement, empowering people to learn about their food system and its cultural dimensions. The challenge for designers today is to develop exciting and innovative proposals for a future Productive City that will capture the imaginations of the public. Urban food production is sometimes eschewed on the grounds that its components are messy and unattractive. For urban agriculture to gain wide acceptance and generate enthusiasm, the design of buildings and the garden spaces around them that incorporate edible landscaping and space for small livestock must be aesthetically pleasing. Designers are uniquely positioned to make a difference in how urban agriculture is perceived in this regard, and this book features designs that are visually striking and artistically engaging by way of example.

In response to the well-documented environmental and social problems generated by industrial food systems, the idea of a "civic agriculture"[18] has been offered with the aim of shortening and strengthening food chains, processing, and distribution services that rely on local resources and serve local markets. Small organizations that rely on local knowledge, contribute to the local economy, and consider the ecological and cultural needs of the community for which they produce food are key to this initiative. Farming expertise must be brought into the planning process as a way to improve city design, to integrate agricultural areas into urban plans and subdivisions, and to ensure viable plots and/or livestock shelters are generated.[19]

Building on the concept of an "everyday urban agriculture," as put forward by Domenic Vitiello and Michael Nairn, Nevin Cohen and Radhika Subramaniam assert that "everyday practices of food production and distribution in cities, the actions of ordinary people in local neighborhoods, register as quiet but persistent challenges to the agro-industrial complex."[20] This leads to questions that this book aims to respond to regarding what it means for individuals in communities engaged in creative practice to reconnect to their food, neighbors, and environment through urban agriculture. The following pages, in short, explore the resulting physical engagement with place that growing food requires. [21]

THE EMERGENCE OF DESIGN STRATEGIES FOR CITIES

As the sizable impacts that climate change, peak oil, and other global shifts have on the provision of food becomes apparent, they demand our attention, and require a greater sophistication in developing sustainable urban settlements in response to broader implications. Questions about net-zero-impact food supply chains for urban residents are directly relevant to the way we design and plan our built environment, and therefore how we educate the designers of the future. The concept of resilient cities, which considers the ability of a city to absorb shocks and changes, suggests a need for more local control of key resources such as food and energy. Urban agriculture can contribute to the future resilience of cities in an era of uncertainty and change.

Not surprisingly, cities are beginning to seriously consider the impact of food on their futures. Urban agriculture and food security have attracted considerable interest in recent years, and lectures, presentations, exhibits, and publications, as well as the number of initiatives reaching implementation have increased significantly. In New York City, FoodWorks is a plan spearheaded by the City Council Speaker to "use New York City's food system to create jobs, improve public health, and protect the environment." The program seeks to create a greener, healthier New York by improving the food cycle of "production, processing, transport, retail, consumption, and post-consumption."[22]

In Rotterdam a civic initiative to develop a food strategy focuses on three ways urban agriculture can be effectively integrated into the conurbation: if it is economically viable, spatially incorporated into the city, and can be woven into the social fabric.[23] Spatial integration addresses the need to find appropriate locations within the city for food production, which may include opportunistic interventions such as the use of niches and overlooked spaces as well as planned interventions that fuse food production systems directly into the design of new buildings. Technical integration addresses opportunities to make connections between productive gardens and available resources such as waste heat, water, nutrients available in waste, etc. Social integration addresses health, community engagement, and education about gardening and nutrition.

Food issues can often, as here, be a point of departure for addressing a range of design considerations including social amenity spaces, cultural contexts, community-centered design, and sustainable building practices. It is now proven that food-related initiatives can lower water treatment costs, reduce the heat-island effect in cities, and cut health costs through better nutrition. These facts have begun to attract political attention, which in turn helps to enable funding for future projects.

IMPLICATIONS FOR DESIGN PRACTICE

A purposeful way to address the sustainability agenda is to develop higher density, multifunctional, and well-integrated communities that reduce the distance needed to transport people as well as goods. A truly sustainable urban community can only be created, however, when the benefits and implications of urban food production, including appropriate building design, urban spaces, and components that can contribute to the local food supply are addressed in a pragmatic way. To do this effectively requires an understanding of the ways in which different urban patterns influence how and where urban residents farm, acquire, and consume food. This understanding will help to conceive new ways to utilize urban spaces and incorporate urban agriculture into a city's fabric as well as its infrastructure and community, while ensuring that it is economically viable.

Productive landscapes do not of course require formal design or planning interventions to be introduced. There are many examples worldwide of everyday urban agriculture that is largely disconnected from the world of professional design,[24] such as the *organopónicos*[25] present in Havana and other Cuban cities, which were created with little intervention from professionals of the built environment. Professional designers, however, are in a position to increase yield or to help residents find previously overlooked productive spaces, creating new types of spaces, buildings and components.

As the projects in this book illustrate, design professionals are increasingly considering the possible synergies between food production and urban design. Whether in the skills they bring to analyzing a physical environment or in the broader processes for designing urban spaces—which are often most effective with resident participation—great potential exists for architects, landscape architects, planners and other designers to work with residents and food experts to transform urban landscapes into productive landscapes. Designers' abilities to offer simple as well as complex design solutions to multiple problems associated with challenging urban sites can provide needed guidance to intensify and expand food production.

In 2005 two important events for urban agriculture considered what designers can bring to questions surrounding urban food systems. The first, a research project at McGill University in Montreal called "Making the Edible Landscape,"[26] launched in four cities on different continents. Second, André Viljoen et al. published *CPULs: Continuous Productive Urban Landscapes*. Both share an integrated and strategic approach to synthesizing the relationship between design and food in the urban context and acknowledge that initiatives offering widespread social benefits can be a challenge to implement in an existing city. The McGill initiative showed that in many cases, in the absence of a coordinated strategy, more opportunistic, small-scale interventions within different zones of the city are often implemented piecemeal. These can, ideally and with the help of designers, eventually be connected into a system such as the CPUL.

Urban areas in Western cities have zones that encompass very distinctive characteristics that offer designers a variety of urban agriculture opportunities. For example:

• City centers often feature dense development due to the high cost of land. In such areas, buildings designed with integrated production components can provide food for residents.
• Traditional urban residential areas that feature a fairly dense urban fabric with smaller yards and limited public green space can focus on strategies that incorporate planters and raised beds, rooftop gardens on new buildings, and farmers' markets.
• Former industrial areas, especially those in transition, offer significant opportunities for urban agriculture. Greenhouses, raised-bed gardening, rooftop gardening, and small agricultural businesses are all suitable for areas of prior soil contamination.
• Older suburbs with larger lots tend to offer enough land for growing in community spaces as well as front and back yards.
• Newly designed, lower-density suburbs often have considerable public land available for food production as well as the front and back yards of private homes.

- Commercial areas at the edge of cities such as business parks and government complexes often have large parking lots and lawns that can be turned to horticulture.
- Periurban agricultural land can be used for extensive food production as well as provide opportunities for integrating recreation and education with agricultural activities.
- Infrastructure—highways, railways, power line corridors—often has considerable adjacent waste land, affording opportunities to integrate the production of food with the spaces provided for energy and transportation.
- A city's solid waste, wastewater, and stormwater can be treated and reused as nutrient sources for crops.
- Sites slated for but awaiting long-term development provide short-term opportunities for community-engaged food activities.

The examples in this book suggest design principles that may help designers and planners continue to innovate as they develop projects that incorporate food. In every instance, the first order of business is to identify local opportunities, such as waste spaces, for integrating food production into existing land use patterns, because these offer the most immediate windows for implementation. Growing surfaces and spaces can also be integrated from the outset into new building projects with advance planning—prime opportunities include roof surfaces, components that allow production on a building's vertical envelope, and sun rooms that double as living and growing spaces. In addition, modular and stand-alone components such as planters can be designed to fit into existing spaces as well as be incorporated into the design of new spaces. Other functions and infrastructure can be used symbiotically, such as the use of waste heat from kitchens to warm greenhouses. As opportunities are identified, both short- to medium-term interventions and long-term solutions should be considered. Most important, engaging the stakeholders through methods such as participatory design initiatives generate useful ideas and feedback from the people who will actually be expected to use the space, and help ensure that the productive spaces will indeed be used.

IMPLICATIONS FOR DESIGN PEDAGOGY

The education of architects, planners and other professionals who impact the design of cities has until recently failed to address the implications of food supply on design.[27] Today architecture, landscape architecture, interior design, urban design, and planning programs are increasingly adding an urban agriculture and food system focus. As these programs expand, current practitioners are being inspired by the imagination of a new generation of students, many of whom are putting what they learn into immediate practice in careers after school, and continuing to inspire others about how urban food security can be addressed in a future certain to include high energy costs and increased planetary temperatures.

Architecture and planning students who have been exploring these questions are trying to find answers to some of the problems many of our cities are facing now, and will continue to face with an increasing amount of urgency, regarding access to food and its relation to sustainable design. These students

have demonstrated that design plays an important role for every profession that seeks to enhance the supply of food to the ever-increasing population of our cities.

The overlapping areas of interest between design and food supply were first formally addressed, identified, and investigated by a handful of architecture programs that have emerged as leaders in this specialty in recent years, including the Architecture Program at the University of Brighton in the United Kingdom and the College of Architecture and Design at Lawrence Technical University in Detroit. Another is Ryerson University's Department of Architectural Science in Toronto, Canada, where students have tackled food issues as design challenges in an assortment of pedagogical contexts, from introductory design studio projects to student-run design competitions and complex thesis projects.[28]

The attention to food issues in these design schools, as elsewhere, has grown out of a number of interests, combining in particular the increasing integration of sustainability and community development into the core curriculum, which of course parallels the rise in environmental and social awareness among students overall. Most do not work on solutions to the problems inherent in today's food supply chain in isolation, but see clearly how they relate to housing, waste management, education, landscape design, and other basic urban spatial demands. Aims that include greening the environment, creating a sense of community, increasing access to fresh produce, and designing in an environmentally responsible way are seen correctly as mutually inclusive. Using food issues as a point of departure has proved to be an excellent way to get them to address a variety of design challenges including social inclusion, cultural context, community design and sustainable building practices. Students themselves led the pairing of food and agriculture with architecture and urban design issues in several of the cases shown here.

Students today, many of whom have grown up with an awareness about sustainability that previous generations lacked, are learning early to see agriculture as an integral part of any city's overarching plan, and are likely to continue to develop this whole-world view as practitioners. They are learning that urban agriculture is a natural partner to adaptive reuse and green building technology projects. These young designers bring a new range of possible solutions to those who advocate for the transformation of food systems. It is through exchanges between professions that innovative, effective solutions can be forged. Where food and urbanism intersect, planning and design practice and pedagogy have the potential to continue to create increasingly successful places for urban agriculture.

IMAGINING THE PRODUCTIVE CITY

The city has historically been a space for food production. Systems that had been in place for centuries, however, gradually weakened and disappeared with the advance of industrialization and modernization. The examples highlighted in this chapter envision reestablishing relations between cities and food production that differ from present conditions. Some emphasize productive spaces as connectors with the potential to unify a city or as solutions for underused land. Others accept the continued expansion of built-up areas into exurban fringes, provided new developments incorporate mixed-use zones for gardening. Cities are also reimagined through a complex interpretation of agriculture that captures the nutrient cycles or intensifies agricultural processes—particularly by rendering farms vertical—to make them adaptable to space constraints. Ultimately, all these alternatives seek to transform barren urban landscapes into edible landscapes.

over 100,000 tons of salad crops annually within the city limits. Residents farmed fully 16 percent of the land with a system that included utilizing the wastes of both humans and horses,[2] but this system declined in the late nineteenth century. It was therefore natural that the visionary city planners of the late nineteenth and early twentieth centuries imagined ideal future cities that, each in its own way, sought to preserve this fecundity and addressed what they recognized as an encroaching threat to rapidly expanding cities: the lack of land reserved or assigned to food production in urban and near-urban spaces. Yet the reality of how cities developed in the twentieth century, particularly in Europe and North America, was fundamentally different, with urban spaces becoming increasingly devoid of food production. As today's decision makers grapple with current environmental, resource, and economic issues, food security and supply is increasingly becoming a significant factor in city planning.

As Carolyn Steel and other writers on urbanism have demonstrated, the relationship between food and the development of cities has long been intertwined. "Food shapes cities, and through them, it moulds us—along with the countryside that feeds us."[1] Even up through the Industrial Revolution, cities had complex food production systems; the Parisian biointensive method of cultivation, for example, achieved extremely productive outputs estimated at

Food Production as a Means to Connect Urban Spaces

Urban agriculture as it is now known tends to be fragmented across a patchwork of productive spaces; André Viljoen and Katrin Bohn of the University of Brighton propose an alternative, Continuous Productive Urban Landscapes (CPULs).[3] They insist on the need for a coherent network of

RIGHT: Master plan for a Continuous Productive Urban Landscape (CPUL) proposed for London; potential productive areas are shaded in green.

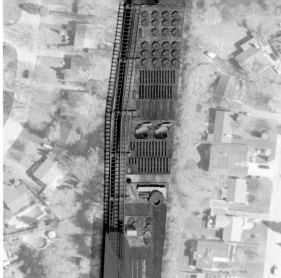

spaces, or "landscape territories that traverse the city,"[4] where food is produced. They propose that for a system of urban food production to become perceived as a whole, it is essential for it to be made up of continuous spaces of cultivation. This approach connects Ebenezer Howard and Frank Lloyd Wright's efforts to link urbanites to their food production systems with the efforts of other urban planners and practitioners, notably Frederick Law Olmsted, who sought to introduce parks systems to North American cities. Viljoen and Bohn show that it is when productive spaces become part of a coherent network that they gain significance and meaning as urban landscape.

Viljoen and Bohn's initial CPUL vision focused on London. The concept's best expression to date is in their 2004 proposal for a "sustainable landscape strategy" for the Thames Gateway, the large redevelopment in East London that is the location of the London 2012 Olympics, and considers the spaces between existing and proposed developments as well as land within the development sites.[5] The design team focused on two zones, developing different strategies for each. For the Lower Lea Valley, a landscape of "hidden treasures" is proposed— small- or medium-scaled productive spaces bundled along the river and canal system with access points at strategic locations. In London Riverside, larger pieces of open land with dramatic river views are strung together by walkable links.[6]

Viljoen and Bohn have continued to develop the CPUL concept, applying it also in the industrial town of Middlesbrough in northern England (see page 26). Other designers have also begun identifying linear urban spaces with particular potential for cultivation. Ravine City/Farm City (see page 30) uses the system of ravines that slices through greater Toronto as corridors for housing as well as food production, and resource conservation. Transportation corridors are also being reworked for simultaneous circulation and cultivation, as in the Post-Carbon Don Mills concept proposed by Ryerson University graduate student

Mike Blois. He explores the potential of a retired railroad track in suburban Toronto, an unsightly divider in the vast suburban expanse, to become a vital link. He is planning for the moment the low-density rings around a city have to confront a post–peak oil scenario, and inhabitants can no longer depend on either the automobile for transportation or long-distance shipping for food supply. He proposes a multilayered system of biking/walking paths and urban farms as a solution.

The potential for interlinked urban agriculture does not depend necessarily on linearity or even physical connectivity, however. Connectivity may also be interpreted as a network built by the establishment of linkages across widely scattered elements of urban agriculture. Bk farmyards, for example, a decentralized food-production network led by architect/urban farmer Stacey Murphy, developed a strategy in 2009 that "uses social media to pool all the resources of Brooklyn into a crowd-sourced decentralized farm." The result is Farmshare, a Web-based platform that allows users to share all resources of urban farming: from donated seedlings grown on a windowsill to a borrowed wheelbarrow for hauling soil.[7] This is but one recent illustration of the potential of electronic communications to foster urban agriculture by creating connectivity among interested parties in cities. Web design or system design plays an essential role in urban design by facilitating the envisioning of all city spaces as potentially productive spaces.

Rehabilitation of Waste Spaces

Even the most dense and dynamic contemporary cities are full of abandoned, underused, or overlooked spaces, as well as land that is adjacent to or under major arterial infrastructure, such as elevated highways and transit lines. Such "junk spaces"[8] or "drosscapes"[9] are often condemned as scars that contribute little to the urban fabric. Their prevalence is also closely associated with poverty levels. A

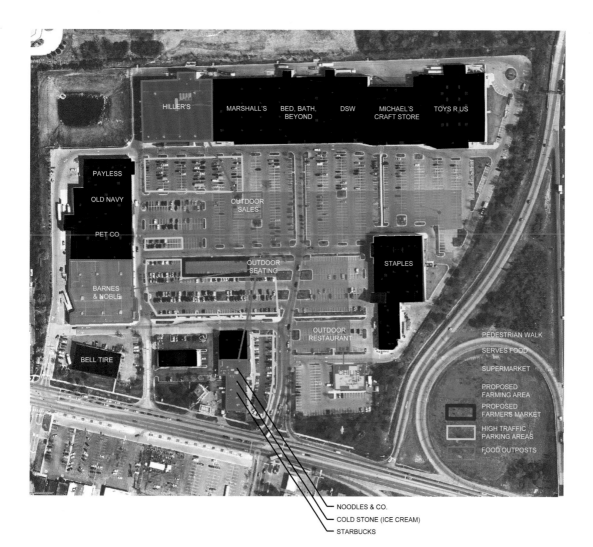

Labels within the image:
HILLER'S MARSHALL'S BED, BATH, BEYOND DSW MICHAEL'S CRAFT STORE TOYS R US
PAYLESS
OLD NAVY
PET CO
OUTDOOR SALES
OUTDOOR SEATING
STAPLES
BARNES & NOBLE
OUTDOOR RESTAURANT
BELL TIRE
CHILI'S
PEDESTRIAN WALK
SERVES FOOD
SUPERMARKET
PROPOSED FARMING AREA
PROPOSED FARMERS MARKET
HIGH TRAFFIC PARKING AREAS
FOOD OUTPOSTS

NOODLES & CO.
COLD STONE (ICE CREAM)
STARBUCKS

ABOVE: Proposal for the Park-N-Farm terraforming concept by Stacey Murphy of bk Farmyards, where underutilized suburban shopping center parking lots are transformed into farming areas.

Brookings Foundation study reported that seventy major American cities averaged 15 percent vacant land area, with higher vacancy rates in the South.[10]

The appropriate response to leftover space is an ongoing topic of debate, with emphasis specifically on how to reduce the impact of essential infrastructure on local neighborhoods. Creative solutions for neglected spaces adjacent to inner-city council estates are being offered in parallel in two British cities: London by the architectural consultants What if: projects, and Birmingham by the municipal initiative GEML.[11] These two different actors have identified, as Jane Jacobs suggested earlier in North America, how gaps within the urban fabric isolate communities, and are developing strategies for appropriating unused spaces to accommodate the needs of the local population, including food.

Abandoned industrial zones, riverbanks, ravines, rail corridors, and even residential backyards can be seen as idle sites awaiting transformation into spaces for food production.[12] Projects such as Andy Guiry's proposal for the Urban Agriculture Hub underneath a waterfront highway (see page 34) illustrate how these spaces can become assets and catalysts for community-based food production. In extreme

instances of urban decay, large expanses of a city may be available for repurposing.

The shrinking of formerly thriving industrial cities is now recognized as a worldwide phenomenon, so much so that a major international research initiative has undertaken a systematic comparison of them.[13] Detroit is often used as the most striking example of a city in decline and many see urban agriculture as key to its rejuvenation. New initiatives—such as a proposal to create the world's largest urban farm, Hantz Farms,[14] on the city's lower east side, and Recovery Park, a planned community redevelopment project that will include urban farming, education, and commercial and housing development[15]—have served as the lightning rods for debate on the best approaches to developing urban agriculture in that city.[16] Hantz Farms represents a pure commercial venture, while Recovery Park devotes its 30-acre site to job training for recovering drug addicts through urban farming. A participatory design process involving multiple partners is crafting the initiative, with the University of Detroit-Mercy's Detroit Collaborative Design Center playing a key role as facilitator. Lawrence Technological University's Detroit Studio has been similarly integrating analysis and design for urban agriculture through its Community Outreach Program for years.[17] Large-scale agriculture projects benefit from lowered land costs and the reduced competition for urban land that cities in decline offer. Agriculture has thus emerged as valuable use of land where the population or the employment base is declining and where built forms are being abandoned.

In flourishing cities, very inventive solutions are required for introducing urban agriculture. Competing land uses mean space is at a premium. There are always overlooked parts of the urban landscape that may be potential resources for food production, however. Bk farmyards's Stacey Murphy saw possibility in shopping centers, where large parking lots are unevenly used. She developed a proposal entitled Park-N-Farm: Terraforming the Strip Mall Parking Lot, whose strategy transforms oversized strip mall parking lots into farms and encourages nearby residents and restaurants to pay for a share of the crop.[18]

The winning entry of the Cities Alive 2009 student design competition, called Cliffside Village in The Post-Carbon City, similarly sought to offer "a viable solution for the revitalization of dilapidated commercial strips throughout North America," making them "the vibrant, pedestrian, and community-friendly environment that [they have] the potential to be."[19] The proposal recommends a rezoning of a commercial site to allow for residential units to be constructed over the existing business strip, as well as "the insertion of green and interactive spaces into the fabric of the site [to promote]"

new and diverse activities, including farmer's markets, playgrounds, urban farming, cycling, etc." Food production spaces are used as part of the strategy for reclaiming car-dominant developments.

Large-scale urban redevelopment is making strides in many vibrant cities as well, often targeting tracts of land that have lost their initial function, including decommissioned bases, derelict port lands, shuttered sanatoria, and especially abandoned factories and warehouses. These are all likely sites for decay and, therefore, reinvestment. Redevelopment plans for such "brownfields" are starting to integrate urban agriculture, especially where they are turned into mixed-use zones. The Southeast False Creek case in Vancouver (see page 44) is one illustration of this trend. Toronto's Downsview Park is also an example of a large-scale development that consists of a partly built-up new neighborhood with a partly open "cultivation campus" (see page 46).

White Bay Eco-City in Sydney, Australia, is a large redevelopment area where an "agricultural precinct" was proposed in 2007 by a team of students from the University of Sydney's Faculty of Architecture, Design and Planning, led by Rafael Pizarro.[20] The team's mandate was to design a carbon-neutral urban zone for 15,000 residents on eighty underdeveloped hectares in Sydney Harbor, near the city's central business district. This agricultural precinct is one of five precincts developed for the city, and it contrasts intentionally with the mixed-use urbanism of the other zones. It features state-of-the-art urban agriculture practices that incorporate water treatment and food production for the entire site, and generates surplus food for surrounding areas.

Some redevelopment projects consider underused lands found in a particular pattern spread across a metropolitan area rather than one large site. The Mayor's Tower Renewal in Toronto reassesses the land surrounding high-rise apartment blocks across the city, whose footprints occupy as little as 10 percent of the lots they rest on. Instead of becoming "towers in a park" as prescribed by the modernist architecture movement that gave them their ultimate form, however, their excess land is rarely used and sometimes fenced off. In addition to improving the buildings, the renewal program seeks to make better use of this enormous, underutilized resource—many apartment buildings occupy fertile land that was under cultivation only a generation earlier. High-rise neighborhoods could thus become "hubs servicing the city at large, with surplus energy sold to adjacent communities and yields from farming initiatives supplying local markets,"[21] and could serve their diverse communities with programming including youth training, seasonal markets, and communal kitchens.

Civic Farm
Steward Farm

Farming Subdivisions

The idea that towns need to be fully integrated with agricultural lands has been seminal since the inception of modern city planning, yet over the past century planners have shunned mixing agricultural uses with housing. Modern farming is regarded as a heavily industrial process that generates noise, odors, pollution, manure, and other wastes and necessitates machinery; conversely, trespass and vandalism can be a problem for farmers.

Recent interest in alternatives to industrial agriculture has led to smaller-scale, organic approaches to food production that are less reliant on chemical fertilizers, insecticides, and herbicides, making dynamic local agriculture—designed to feed local people—able to be combined successfully with residential land uses in periurban areas. The propinquity of home and farm is now also seen as a learning and economic opportunity for suburban residents. The loss of farmland at the edge of conurbations is well documented and of continuing concern, and efforts at preserving traditional farmland have been costly and only moderately successful. New creative ways to maintain food production while enabling some land development have evolved as a balance between population growth and agricultural preservation.

One successful model in North America has been the adoption of conservation subdivisions that accommodate the pressure for new residential development allowed under zoning regulations while protecting agricultural land and natural resources.[22] Usually long-term agreements with farmers in the community encourage or require adoption of organic and small-scale farming practices to minimize impact on the land and on neighbors and provide jobs within the community. The accepted uses need to be clearly outlined and understood by both farmers and residents to maintain good neighbor relations. Typically conservation subdivisions are protected through easements or land donations that limit what type of development is allowed.

Smaller housing lots arranged in clusters free up land on development sites for food production, forestry, wastewater disposal, or outdoor recreation. These propose "mixed land-use garden-zones [that] are conceptualized as . . . a hybrid zone where the mixing of the urban and agricultural is encouraged."[23] Typically, at least 40 percent of the land is set aside. This arrangement also reduces the need for infrastructure—such as streets and sewer lines—and provides financial and environmental benefits for the developer and the municipality. For example, the developer of the 418-acre Farmview subdivision in Pennsylvania donated 145 acres of cropland to a local conservation organization. The land was separated from the houses with a thickly planted area 75 feet (25 meters) deep. The resulting half-acre residential lots proved popular due to their proximity to this amenity despite being considerably smaller than typical 1-acre lots in the area.[24]

Farming subdivisions have evolved in recent years from conservation subdivisions. Here residential uses are mixed with working agricultural land, such as orchards, vineyards, annual and perennial crops, or livestock. Typically these developments are 40 acres or larger. Currently under development, the Agritopia Project[25] near Phoenix, Arizona, is a 210-acre mixed-use community centered on a local farm owned by the Johnston family. Specialty organic crops are sold at a farm stand on the site and used at a local restaurant. The zoning was arranged to provide space for gardens, stores, educational facilities, restaurants, and community buildings, and, most important, more than one dwelling unit per lot. This flexibility helps adapt lots as family needs change—family gardens, second units, or small business facilities can be carved out of each space. Community farm lots of 400 square feet (37 square meters) in raised beds are also available for community members to rent.

Colorado-based consultant TSR Group has proposed the "Agriburbia" concept, which intersperses homes with farms and gardens.[26] Platte River Village, a 618-acre site in Milliken, near Denver, Colorado, is in development now and includes 944 planned homes surrounded by 108 acres of backyard farms and 152 acres of drip-irrigated community farms.[27] Arguing that agriculture needs to become basic infrastructure for new subdivisions, TSR's founder, Quint Redmond, split the types of farms within Agriburbia into two groups: Civic Farms, professionally managed by farm contract on publicly or part-publicly owned land, and Steward Farms, which help landowning residents participate in food production with the support of professionally managed farm services, either on their own land or on privately held lots within the development area.[28]

North American developers are starting to see farms as an amenity that adds value to residents. The Urban Land Institute estimates that there are currently at least 200 projects in the United States that include agriculture as a key community component.[29] While many of these are aimed at affluent homeowners and focus on specialty agriculture such as vineyards, some are beginning to integrate more diversity. Prairie Crossing (see page 50) is an example of a farm subdivision that addresses a variety of environmental and community issues, including resident backgrounds.

Farming subdivisions, however, can have the unexpected consequence of exacerbating sprawl at the edge of cities. Many of these developments also have limited community diversity and consist of large family houses in places that are poorly served by public transport, requiring the residents to rely

RIGHT: Winning entry in the Cities Alive 2009 competition, showing the many ways a strip mall could be transformed to include productive spaces.

on private cars for transport and thereby offsetting the environmental benefit of reducing food miles by growing crops close to home. When designed close to transportation hubs and as mixed-use communities with employment potential, they can reduce the overall environmental footprint of their residents. Treasure Island is an interesting example of successful adaptation. This development[30] on a 400-acre former naval base on an island in San Francisco Bay integrates organic agriculture into an 8,000-resident neighborhood where all the homes are within an easy walk of a new ferry terminal. Plans include repurposing of an old air hangar as a market hall and artisanal food production space.

In the future, developers must consider how principles from farming subdivisions can be integrated into existing urban areas, and they must address affordability, diversity, and access to public transport. Many existing suburbs have considerable areas of underused land with food production potential, but land ownership issues preclude easy integration of farming subdivision principles.

Closing the City's Nutrient Cycles

One of the most compelling reasons for integrating agriculture and urban settlements goes beyond the provision of food sources alone—it is to restore the fundamental place of the food system in the broader urban nutrient cycle. A complete or ecologically sustainable design for a city would be a closed loop, with all the wastes from one process used as an input to another process. Urban

agriculture has a large role to play in closing open, polluting loops in the nutrient cycle.[31] Designers are now beginning to propose built-environment patterns that address the urban food cycle and offer responses to the problems in urban resource and nutrient cycles. Through their efforts, food production, processing of food wastes, transformation of urban solid and liquid wastes, energy generation and conservation, and other goals of environmental enhancement are combined in highly creative ways with the conception of environments for living, working, playing, and circulating.

The Netherlands's Innovation Network offers two cases that show approaches to closing the urban nutrient cycle. Agroparks (see page 54) is a general concept that proposes solutions for different geographic contexts, from urban industrial estates to rural zones close to a city. It relies on clustering agricultural functions together to close resource and waste flows. Greenhouse Village (see page 41) responds to the consumption patterns of Dutch flower production in greenhouses and proposes a form of urbanization that captures the resource use associated with this specific industry. Examples in other chapters, such as the Center for Urban Agriculture, Seattle (see page 144), the Science Barge, Yonkers (see page 86), and Agro-Housing, Wuhan, China (see page 140), apply similar principles at a smaller scale.

These examples are clearly based on high-technology solutions that accept industrial agriculture. Other approaches seek the same general goals but adopt simpler agricultural techniques, often referred

to as "appropriate technology." There is indeed an ongoing debate within the urban agriculture movement regarding the tenets that should govern the movement—a central question is how best to connect the food system to other nutrient systems in the city. The choices available result in a variety of physical implications.

Vertical Farming

One of the key questions related to urban agriculture over the past decade has been whether to focus on developing vertical or horizontal designs. High-rise farms have attracted considerable attention recently, particularly following the pioneering work of microbiologist Dickson Despommier, a professor of public health at Columbia University. His Vertical Farm Project for New York grew out of the realization that as world population grows there may be a shortage of suitable horizontal surfaces close to urban areas with the capacity to produce the large quantities of food needed to feed cities. Despommier used data from NASA to estimate that each person could be fed from the produce of 300 square feet (30 square meters) of intensively farmed land using current technologies, and he speculated that this area could be reduced as technology progresses. His team calculated that an urban farm the area of a New York City

block and thirty stories in height would provide about 3 million square feet (280,000 square meters) of agricultural space—enough to feed 10,000 people.[32] Potential benefits would include zero need for food transportation, on-site waste recycling, a year-round food supply with reduced reliance on weather, and economic benefits for the city. The design potential of these ideas has intrigued many architects, and a number have collaborated with him to give them physical form. One example, the Pyramid Vertical Farm conceptual design by Despommier and Eric Ellingsen, presents an architectural interpretation of a vertical farm for Dubai.

The vertical farm concept has inspired many other architects, as well as futurists and speculative capitalists, to develop their own visions for productive high-rises around the world.[33] The Plantagon Greenhouse is a Swiss-American proposal based on an idea by Swedish ecofarming innovator Åke Olsson for a massive 100-meter-high glass structure that features a spiral platform for plants.[34] Crops are planted and placed on the bottom of the spiral—which is synchronized with the plant's development cycle—and it rotates, moving them up the inside of the structure as they grow. Crops are harvested when they reach the top. The Plantagon project is expected to cost about $100 million to build, and a feasibility study suggests that the largest version of the structure would take three years to recoup its initial investment, while smaller versions would take about ten.

Another creative project is the Dragon Fly, a dramatic 600-meter-tall vertical farm shaped like the insect's wing, designed by Belgian architect Vincent Callebaut. This greenhouse of glass and steel is proposed for Roosevelt Island in New York.[35] It would be heated by solar energy and cooled with passive ventilation. Water would be collected from exposed gardens and treated for use on interior farm space. Urban farming space for raising cattle and poultry and twenty-eight different types of crops would be spread over 132 stories.

Architecture students have been particularly attracted to vertical farming for its combination of high technology, with the use of sustainable design to address challenges in food systems locally and globally. Some designers have also used vertical farming proposals as an imaginative instrument to convey the dangers or negative effects of current industrialized agriculture practices in ways that are easy for the general public to visualize and quantify. One such case is Pig City (see page 56), a provocative representation by architectural firm MVRDV that explores the prevalence of pig-farming operations and national pork consumption in The Netherlands. These idealized proposals, whether sincere or self-reflective, generally rely on established technologies, such as advanced greenhouse systems, hydroponics, aeroponics, passive solar systems, and biogas energy generation, but combine them in creative though untried ways. The vertical farm concept has therefore been criticized for its complexity and expense as having little practical significance for addressing world food problems.

The most significant critique of these proposals is the problem of accessing the light necessary for plant growth. Stories of crops stacked on top of one another reduce access to daylight on a structure's interior, and so, even for structures based on all-glass curtain walls, significant amounts of electricity for artificial grow lights would be necessary to make expected agricultural yields materialize. Stan Cox and David Van Tassel estimate that for the United States to grow 15 percent of its annual wheat crop vertically, would require all the electricity the country generates in one year. The food would have a very high energy input per calorie of useful food energy.[36] Critics claim therefore that even if vertical farming were feasible on a large scale, it would merely extend the dependence of food production on current industrial mechanisms, chemical fertilizers, and fossil fuels, thus removing agriculture even further from natural systems.

Conclusion

As urban agriculture's potential for promoting community, social, economic, and environmental goals has gained recognition in recent years, it is increasingly being integrated into large-scale development projects by architects and developers, like the winning design for the Low2No design competition to build a sustainable building complex on a reclaimed harbor in Finland.[37] Furthermore, food production has moved up the municipal agenda, so city authorities in many countries have begun to take increasingly strong interest in it. Partly through increasing awareness by planners and partly through pressures from citizens,[38] urban agriculture is now being considered in many comprehensive and neighborhood plans in North America. For example Seattle's 2005 comprehensive zoning plan incorporated at least one community garden for every 2,500 households.[39] In March 2010, a report submitted to the Detroit City Council by the City Planning Commission similarly suggests the drafting of a policy and a zoning code that are specifically focused on allowing and facilitating urban agriculture in the city.[40]

Appropriate municipal support is crucial to the wider adoption and success of urban agriculture. Some cities assist through zoning policies, land allocations, strategy development, and even funding. Local policies are often still based on outdated planning and community health attitudes that can present significant barriers to food production, however. Bylaws that forbid the keeping of chickens in many urban areas represent one example of bias against urban agriculture. Happily more municipalities have begun to recognize agriculture as a basic use of urban land and look for opportunities to integrate it into urban and regional planning. Policies favoring urban agriculture can have many implications for the physical form of cities as they develop over time. Specific urban design approaches are increasingly understood as either hampering or supporting the continued development of urban agriculture, and architects, landscape architects, and other designers are finally advocating moving from merely reimagining productive cities to designing functional and attractive productive cities and, ultimately, to realizing their designs.

RIGHT: A rendering of the proposed Plantagon greenhouse.

MIDDLESBROUGH URBAN FARMING PROJECT

DOTT07, JOHN THACKARA, DAVID BARRIE, ZEST INNOVATION, DEBRA SOLOMON, BOHN & VILJOEN
MIDDLESBROUGH, UNITED KINGDOM

The urban landscape of Middlesbrough, a former industrial town in the United Kingdom, was radically transformed through a design initiative called Middlesbrough Urban Farming Project, a part of the Designs of the Time 2007 (DOTT 07) yearlong series of community projects in the northeast of England that explored how design could improve lives in meaningful ways.[41] DOTT 07 focused on sustainable alternatives for living, exploring several themes, or "zones:" energy; schools and schooling; health; and food, which was called the "ultimate design challenge."[42] Middlesbrough became a living laboratory for the food zone, an example of how urban agriculture can be integrated into cities through design interventions and how healthy food can become woven into their basic fabric.

The long process of urban transformation began in October 2006, when David Barrie, a senior producer for DOTT 07, and his team consulted with community groups and organizations in Middlesbrough. His project-based approach included participation from many disciplines and interest groups, and created a groundswell of support for what they called a "soil-to-table" project.

Collaboration between the Middlesbrough Council, the design team, and well over 1,000 local participants—with additional support from many local and government agencies and organizations—made the vision become a reality.[43] A design team composed of everyone from artists to communications specialists worked closely with the various groups to turn ideas into physical designs.[44]

Architects Katrin Bohn and André Viljoen, who had already developed a plan of Continuous Productive Urban Landscapes for London that creatively uses available urban land for food production, were called upon to map the collaborative design that ultimately identified eighty locations for productive growing.[45] The mapping project, called "Opportunities for a Green and Edible Middlesbrough," was a comprehensive record of places for urban food production, revealing existing allotment gardens, where residents indicated a desire to grow food, and other potential productive locations. Once connected these spaces show an emerging green network across the town. The map also identified which major streets could be lined with productive plants, parks that would be suitable for allotment gardens and orchards, and the schoolyards, laneways, balconies, and other underused open space that could become part of an urban farming initiative.

In May 2007, planting began all over the city. DOTT 07 distributed seventy-two window boxes—sixty-eight medium containers, 1 meter square; twenty-four large containers, 4 meters square; and seventy barrels—to citizens. Two kitchen gardens were also established, one at the art museum and one in the city's main park. Community groups, volunteer organizations, youth and school groups, public health organizations, and even preschools organized to grow food in leftover and adaptable spaces, enacting the design ideas shown on the map. Knowledgeable allotment gardeners, farmers, horticulturists, and the nonprofit organization Groundwork South Tees supported inexperienced gardeners. Through this and other growing initiatives, some developed even without the formal support of DOTT 07, formerly bare back streets have become green and productive.

When the crops matured, harvested food was brought to a public food preparation activity at a playground. Finally, in September of that year, a huge outdoor town banquet dubbed "Meal for Middlesbrough" celebrated the harvest, with over 1,500 citizens eating the fruits of their labor. Additional baskets of food were then brought to the main DOTT 07 festival location at Newcastle upon Tyne, where the landscape mapping project was on display, further educating people about the urban design initiative. The design festival brought international publicity to Middlesbrough for the successful realization of this productive landscape.

Subsequent town meetings in 2008 discussed how to improve the initiative by increasing the variety of seeds available and providing support for growing challenges, such as identifying suitable crops for school groups. Momentum increased, and by March 2008, existing allotment

gardens had waiting lists over 100 people long. At last count in 2010, over 200 community groups were involved in growing food, and three community orchards had been planted with fruit and nut trees, berry plants, and rhubarb.

Support networks established as part of Middlesbrough's "Healthy Town Initiative" enabled the implementation of designs for the continued cultivation of productive landscapes. Because of this, Middlesbrough's experiment has a legacy that remains today. In 2008 Council Renewal Officer Ian Collingwood stated, "This has caught people's imagination. But we've gone beyond novelty now and people want to make it a mainstream activity."[46]

The annual Middlesbrough Town Meal has taken place for four years running. It continues to increase awareness about food miles, sustainable development, the pleasure of eating local produce, fresh fruits and vegetables, public health, and active lifestyle and eating issues related to nutrition and obesity. The initiative has lead to £4.1 million in grants from the government's Department of Health Healthy Community Challenge Fund, underscoring the opportunities presented by concentrating on a city's full landscape as part of a sustainable agenda for designing the built environment.

Design strategies implemented in Middlesbrough, including community involvement from the outset, providing a variety of containers sized for different conditions and plots, and using mapping as an analytical tool for revealing the strategies and opportunities for urban food production, are good examples other cities and towns can follow. The next stage in the evolution of urban agriculture is to incorporate productive gardening from the very outset of any urban design project. As Viljoen explains, "We need to start thinking about food production being part of a city's infrastructure—like roads."[47]

TOP: The plan for Middlesbrough's CPUL on display at DOTT 07.
ABOVE: Vegetables growing in containers provided by DOTT 07.
OPPOSITE: A Middlesbrough resident working in one of the many community gardening plots scattered throughout the town.

ABOVE: A preparatory sketch for the mapping project that identified green corridors throughout the city.
RIGHT AND BELOW: Plan showing the overall Continuous Productive Urban Landscape for Middlesbrough that identifies over eighty sites for growing produce across the town.

An edible Middlesbrough

Middlesbrough CPUL

What if more land in our towns and cities were given over to 'edible landscapes'? The raised green panels show how a network of spaces for growing food, circulation and leisure could be introduced in to the town in the future. These spaces would incorporate market gardens for growing fruit and vegetables and could form part of the town's network of open urban spaces. We call this network of spaces a "continuous productive urban landscape" or CPUL. CPULs are a way to enhance the urban environment and reduce its ecological footprint.

Middlesbrough today

Allotments
There are already many allotments in Middlesbrough. They show that the town already has an infrastructure of urban agriculture. In the future, allotments could become an essential part of an extended network of 'edible landscapes' that run through the town.

The DOTT07 urban farming project

Small containers
142 Window boxes and barrels were distributed as containers for food to be grown in by individuals and organizations.

Medium containers
68 medium containers, one metre square in area, were used for growing fruit and vegetables. These were looked after by schools, community organizations, hospitals and amenity groups.

Large containers
48 large containers, two metres square, were cultivated by schools, neighbourhood centres and other local organizations.

Food was also grown in the town's parks and open spaces by the horticulture department of the local authority and the town's principal art gallery - the Middlesbrough Institute of Modern Art (MIMA).

Across the growing season, Middlesbrough's new 'urban farmers' harvested and ate food they had grown and the final harvest yielded a bumper crop that was shared by over 2500 people in a celebratory town meal.

Dott 07 Opportunities
for a green and edible
Middlesbrough

01 An urban design concept
* plant continuous open space corridors (CPUL) thereby connecting the city with the rural, the wild
** benefit from this new landscape productively in a variety of ways :

02 movement
* improve non-vehicular movement and access by foot or bike throughout the entire town
** reroute traffic

03 energy (economics)
* use the ground more effectively in economic terms, esp. through new types of urban farming sites
** provide employment and invigorate districts through productive elements of the new landscape

04 school
* offset the building density with extra large open space to provide children with healthy and self-sufficient activity options
** improve safety for children with play space weaving through their town

05 health
* offset industrial/noise pollution with contrasting calming and oxygenising open space
** improve air flow in and out of the city through open corridors

06 food
* plant urban agriculture sites in the heart of the town producing organic and local food
** improve the sense of place, the food and eating culture by providing space for food production and processing

07 An urban lifestyle
* preserve the greenbelt by offering the rural on the urban doorstep (within a CPUL)
** enhance people's relationship with and enjoyment of nature, the year's seasons and weather

The DOTT 07 Urban Farming Project
in
Middlesbrough

represents the first practical testing of a concept for continuous productive urban landscape (CPUL). Individuals and organisations participated by growing fruit and vegetables in small, medium and large containers. Over 200 containers were distributed across the city. There was and is a positive acceptance and enthusiasm for urban farming, evidenced by the number of participants who wish to continue growing fruit and vegetables next year and several who wish to expand the area under cultivation. People enjoy being close to edible landscapes.

When imagining how Middlesbrough may develop the CPUL concept in the future, it is important to realize that it does not require everyone to grow their own food. It rather proposes that commercially viable market gardens would form part of the city's network of open urban spaces. In this way, the city would significantly reduce its ecological footprint while at the same time enhancing its urban environment. CPUL provides more experience with less consumption.

An edible Middlesbrough

RAVINE CITY

CHRIS HARDWICKE AND HAI HO
TORONTO, ONTARIO, CANADA

Toronto's natural ravine system is one of the city's defining physical characteristics. Green corridors house creeks and rivers that run directly through its heart. As the city expanded in the nineteenth century, the ravines were often seen as an impediment to creating a gridded street plan and therefore as an undesirable feature. Many ravines were filled in to support the construction of roads and buildings, while others were engineered to act as flood-relief channels or as transport corridors, reflecting the prevailing attitude toward nature as something to be subdued. Today only the largest ravines, such as those housing the Don River, the Humber River, and the Rouge River, remain, and serve as parks and open spaces for recreational activities. They continue to be disconnected from everyday city life, however, and are often dominated by rail lines or highways.

Another notable feature of present-day Toronto is its ecological footprint—calculated at about 200 times its actual land surface area, with almost a third of this attributable to food supply. Despite being surrounded by nutrient-rich agricultural land, only about 20 percent of Toronto's food supply is sourced locally. The remainder is shipped in from distant farms, and the prospects for agriculture on lands adjacent to the city proper are being continually encroached upon by urban sprawl. The 1,200-kilometer ravine system, however, offers large corridors of green space that have significant potential to be exploited for urban agriculture.

Ravine City is a city-scale vision that uses the natural infrastructure of the continuous watersheds and ecosystems of Toronto as infrastructure to redefine it. This proposal by local designers Chris Hardwicke and Hai Ho advocates for an urban system of collective housing and food-producing gardens that are connected to Toronto's ravines and rivers. Through it, they seek to renew the city's ties to nature while allowing densification and further development within the conurbation. Ravine City proposes to restore some of the lost ravines through excavation and remedial demolition. The new ravines would be integrated into the city's infrastructure, including energy, water, waste, and food systems, but would function much like the natural ravines in

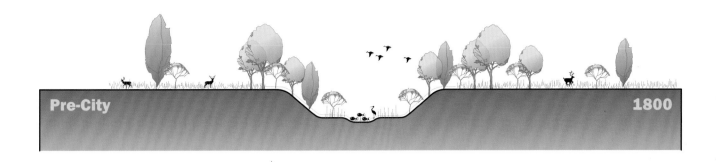

Pre-City   1800

Victorian City   1900

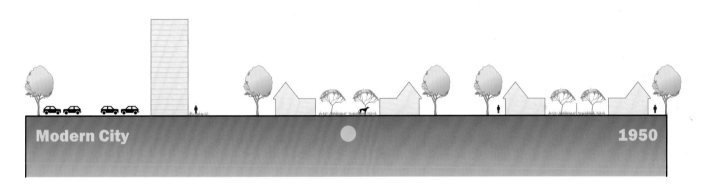

Modern City   1950

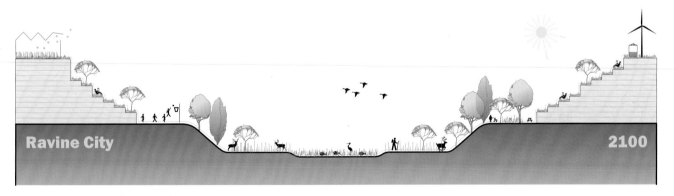

Ravine City   2100

ABOVE: The evolution of Toronto's ravines as they are gradually filled in over time; Ravine City proposes restoring them to their original paths but with new adjacent development.
OPPOSITE: Ravine City's proposed housing units cascading down the topography's natural slope; productive roofs grace each unit. The downtown Toronto skyline is in the background.

terms of controlling water flow and regeneration, cleaning the air, and creating habitat and biomass. Maintained and operated by the city, this new topographic infrastructure would connect with the natural ravine system and operate as additional public space.

Ravine City buildings are designed to integrate opportunities for food production in productive gardens on roofs and terraces, allowing urban dwellers and communities to take responsibility for the production, processing, and delivery of their own food. Existing housing blocks along the ravine's edge would be connected to the green corridors via new podiums and gardens, proposed residential high-rises incorporate vertical growing spaces, and densely planned new housing would hug the ravine's edge. Open land in the existing and re-created ravines is also dedicated to food production in the proposal.

By grouping housing and agriculture in the same buildings, Ravine City creates symbiotic relationships between food, energy, water, and waste. Solar energy generated from the large glazed surfaces of greenhouses is used to create heat for both crops and housing units. Solar panels also collect heat and electricity for use in the buildings. Thermal energy is stored below ground, in geothermal boreholes. Agricultural activities from greenhouses and green roofs create biomass that can be used for energy and compost. Gray water and compost generated by inhabitants replenishes the soil in roof gardens and greenhouses. The concepts employed here are based on closed-loop systems where wastes from one process become a resource for another.

Roofs and facades are increasingly recognized as potentially useful surfaces for food production, energy generation, and water harvesting. Roofs are exposed to sun and rain, and in most conventional roofing systems this is seen as a problem, but for urban agriculture it is a benefit. Here, these features are also utilized—the design of the housing developments creates a continuous, connected growing system along the top edge of the ravines, based on terraces that provide a water collection and management system. The productive capacity of the roof surface becomes a dynamic part of the urban fabric, linking these spaces to the infrastructure systems and ecological networks of the city.

In Ravine City, individual residences have overlapping functions that maximize the potential of their location and use and contribute to the health of the community. Some buildings function predominantly as solar generators; others catch wind energy, produce food, treat wastewater, collect storm water, or provide recreational space. This project proposes a communal vision in which each unit shares its surplus of food, energy, or nutrients, and the units work together to create a flexible and resilient urban network. By creating living and growing space in a dense vertical format, Ravine City aims to reduce the need for sprawling suburbs, eliminating food travel distance and creating a living architecture that is part of a city ecosystem that encourages urban dwellers to take more control over their own ecological footprint.

BELOW: A view from the proposed terrace roof gardens toward the ravine floor.
OPPOSITE, ABOVE: Sections of Ravine City housing showing the relationship of productive crops to ornamental shade trees.
OPPOSITE, BELOW: In Ravine City, energy, water, food, and waste systems are integrated.

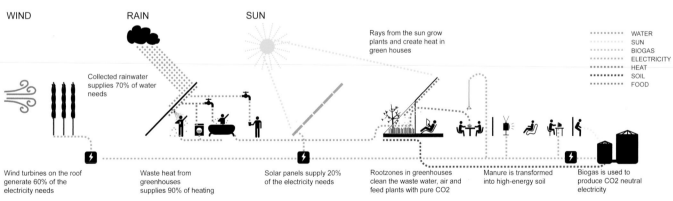

WIND   RAIN   SUN

Rays from the sun grow
plants and create heat in
green houses

········· WATER
········· SUN
········· BIOGAS
········· ELECTRICITY
········· HEAT
▪▪▪▪▪▪▪▪▪ SOIL
········· FOOD

Collected rainwater
supplies 70% of water
needs

Wind turbines on the roof
generate 60% of the
electricity needs

Waste heat from
greenhouses
supplies 90% of heating

Solar panels supply 20%
of the electricity needs

Rootzones in greenhouses
clean the waste water, air and
feed plants with pure CO2

Manure is transformed
into high-energy soil

Biogas is used to
produce CO2 neutral
electricity

GARDINER URBAN AGRICULTURE HUB
ANDY GUIRY
TORONTO, ONTARIO, CANADA

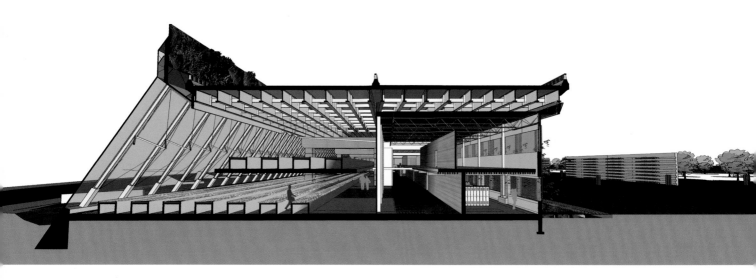

This theoretical project explores the potential for food-related activities to occur at the intersections between different elements of linear urban infrastructure, such as elevated highways, rail lines, pedestrian and cycling trails, and arterial roads. These unused or underused, marginal landscapes, sometimes referred to as "junk spaces" or "drosscapes," are often located in areas that are home to low-income residents and can themselves further localize social and economic poverty. They are frequently condemned as urban scars, disconnected from and destructive to their natural context. Many cities around the world are currently addressing appropriate responses to such wasted spaces and, specifically, how to reduce the impact of essential infrastructure on local neighborhoods. This project was conceived as an intervention that embraces the history of a place and uses agriculture and design to propose new relationships between urban development, infrastructure, and natural systems.

The Gardiner Urban Agriculture Hub is proposed for a site in Toronto, Canada, under the Gardiner Expressway where the Don River enters Lake Ontario. The heavily contaminated site was once home to an oil refinery and consists of waste land along and below the elevated highway. The aim of the proposal is to knit social, economic, and ecological processes with essential urban infrastructure and turn the waste landscape into an agricultural resource. It also proposes architecture that embodies the new potential for urban food production, education, and research. The full project consists of both a greenhouse under the Gardiner Expressway and numerous site-improving interventions. In addition to salvaging neglected land, the structure is intended to reuse components and recycled materials available on-site or nearby wherever possible. Site interventions flow from the principles and practices of sustainable agriculture and sustainable architecture, and are based on ecological processes specific to time and place.

The project utilizes the physical characteristics of the elevated expressway and the space beneath it to create a greenhouse for food production. Since the highway runs from east to west, the area below enjoys considerable solar exposure, and the greenhouse was designed with a south-facing glass elevation that would harness it. The greenhouse also

benefits from the considerable thermal mass of the existing concrete highway structure to regulate temperature inside the building. Educational facilities, food processing and storage, and a commercial space that acts as a local garden center were integrated into the north side of the structure.

The Gardiner highway's structure is referenced in the site's southern landscape through rammed-earth benches that act as visual dividers for a semiwild orchard, which is planted with apple seeds. The chance growth of the apple orchard is contrasted by fruit trees planted on the median of the adjacent road, which are subject to the human manipulation of tree growth known as "pleaching"—a process whereby trees are woven together and fused into living sculptures. These trees align with the rammed-earth benches and echo the rhythm of the Gardiner.

Ecological processes formerly repressed by engineering, such as the

TOP: Community gardens and remediative landscape planned for the area north of the highway.
ABOVE: Plants incorporated into a rammed-earth wall are designed to remove contaminants in the site's soil and to simultaneously contribute to the wall's gradual demolition until it becomes a mound of fertile soil that can be spread over the site again.
OPPOSITE: A cross-section of the Urban Agriculture Hub; the expressway itself functions as the roof of the structure; a greenhouse and programming spaces occupy the area underneath.

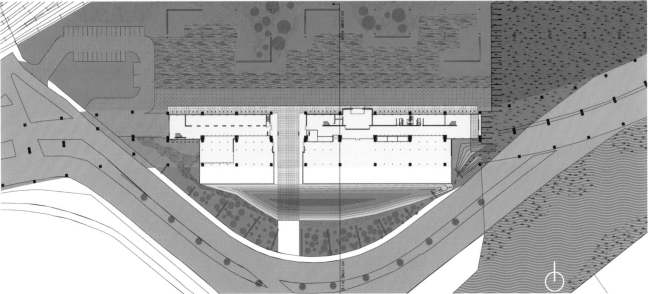

TOP: The view from the Hub's south facade showing the long, glass curtain wall of the greenhouse designed for maximum solar exposure and the rammed-earth benches that separate sections of a proposed orchard. The orchard's appearance in each of the four seasons is indicated from spring, at left, to winter, at right.

ABOVE: Site plan showing how the Urban Agriculture Hub fits into a parcel of underused land beneath the Gardiner Expressway in Toronto.

seasonal flooding of the Don River, are allowed to overtake the site again and are integrated into a hydrological cycle that includes the collection and remediation of water from the Gardiner. This is reused in the greenhouses and surrounding productive landscapes. Remediation of contaminated soil is creatively addressed in the landscape to the north of the building with walls built from highly compressed contaminated soils that would contain specific plant materials able to remove or transform the contaminants in the soil while simultaneously eroding the walls slowly. Over time, what was once a highly structured and contaminated landscape would become a renaturalized park space with soil suitable to support productive community gardens.

The north side of the building itself also deals with localized air pollution from the highway above. The air intakes for the building are below densely planted columns of vegetation held aloft by reused chain-link fence. Vegetation filters much of the particulate matter out of the air before it enters the building and also enriches the air with oxygen. Mini-turbines placed on the median strip along the highway generate electrical energy for the building from the wind created by passing cars. This, combined with a green wall "billboard" running the length of the greenhouse thermal stacks, creates a visible presence to the cars passing above.

The project raises questions about the nature of human control over the environment. Imposed order and chaos are used to contrast human interventions within a remediated, renaturalized landscape. The Gardiner Hub addresses many issues, including the potential offered by waste spaces within a city for creating unique architectural solutions that enable local food production and remove urban blight through sustainable design solutions. By identifying and utilizing the positive characteristics and opportunities available below and around raised highways and on other unused or underused land, it points the way for similar concepts to be implemented on other urban waste land and suggests a solution for reuniting nature with urban infrastructure.

MAKING THE EDIBLE LANDSCAPE
MINIMUM COST HOUSING GROUP AT MCGILL UNIVERSITY, INTERNATIONAL
NETWORK OF RESOURCE CENTRES ON URBAN AGRICULTURE AND FOOD SECURITY,
ETC-URBAN AGRICULTURE
ROSARIO, ARGENTINA

Finding food can be a daily struggle for marginalized urban dwellers. In the informal settlements on the edges of urban regions in developing countries, this problem is partially solved with urban agriculture. Until recently such strategies for combatting hunger and malnutrition have not been seen as an integral part of the infrastructure or capacity building programs that improve the quality of life for residents. Urban agriculture can also provide an economic boost to low-income people by providing additional earning power to urban vegetable growers who sell crops. Urban landscapes can be designed to provide areas for food production as well as other beneficial public amenities at a very low cost. Participatory design processes have been proven to help stakeholders feel an important sense of ownership in a neighborhood, and directly impact the successful cultivation and maintenance of productive spaces.

The Minimum Cost Housing Group at McGill University in Montreal, Canada; the International Network of Resource Centres on Urban Agriculture and Food Security (RUAF);[48] and ETC-Urban Agriculture, based in The Netherlands, began the Edible Landscape Project in 2002. Rosario, Argentina, was one of four sites where edible landscaping strategies were developed—the others included Colombo, Sri Lanka; Kampala, Uganda; and Montreal, Canada.

Rosario, the third-largest city in Argentina, has a population of over one million and high unemployment that increases every day as migrants from both urban and rural areas settle in its informal communities, where they have problems finding jobs, decent housing, and adequate nutrition. Two of these settlements were chosen for this project, the neighborhoods of La Lagunita and Molino Blanco Sur. La Lagunita residents had never tended urban agriculture plots, but residents of Molino Blanco Sur had a history of tending local plots, which gave rise to different needs for each.

In Rosario, the Edible Landscape Project team observed numerous vacant properties that totaled about 35 percent of the land. Ordinances that easily allowed the conversion of unused public and private land to productive gardening were already in place, creating an opportunity to use the wasted space for urban agriculture.[49] Some of the land was already being cultivated, so this project built upon existing informal and formal initiatives. The team developed a participatory design process in which planners, architects, urban agriculture specialists, and residents worked together to envision and design a productive landscape. The consensus-building aspect of participatory design was seen as an essential part of increasing the likelihood of success for the project.

Preparatory fieldwork for the design workshops included interviews with important neighborhood figures and

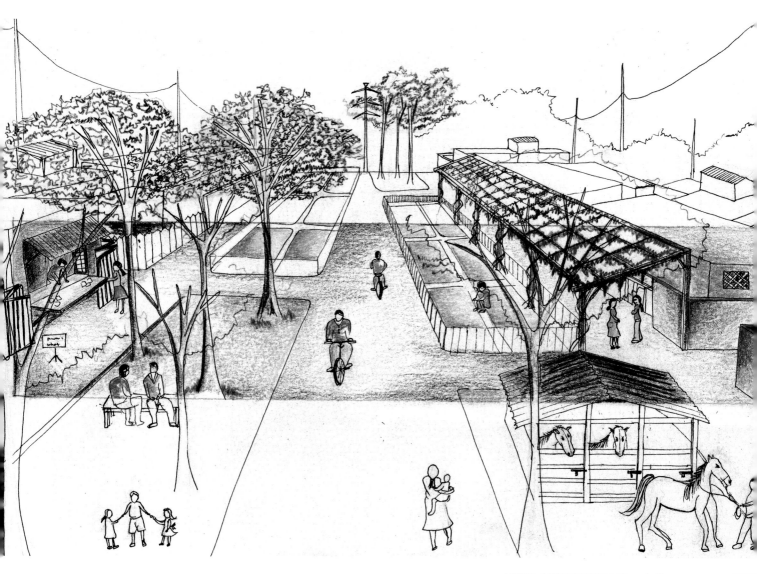

ABOVE: A street featuring both productive landscapes, on the right, and a market for selling locally grown produce, on the left.
OPPOSITE: Inviting and participatory neighborhoods focused on urban agriculture feature productive plots in front of private homes and on sidewalks, adequate lighting, shade trees, and informational boards that encourage community participation.

TOP, ABOVE: Mounds of compost that enable organic gardening.
CENTER, ABOVE: Community members planting Rosario's new fields.
ABOVE: Market day in Rosario; residents purchase locally grown vegetables, giving the low-income families that grew them an economic boost.

stakeholders as well as meetings with municipal and state agricultural officials and interest groups. The team prepared maps, photos, diagrams, videos, and models to facilitate discussion and design, and included examples of similar successful solutions implemented elsewhere to encourage the citizens to see potential. Designers, planners, and agriculture specialists introduced participants to several typologies that would efficiently use land and resources for production and other beneficial activities: garden-parks that combine leisure activities with productive gardens; educational-productive squares that combine cultural events, the education of schoolchildren, productive plots, and sometimes commercial uses; and productive streets, small-scale interventions incorporating market stalls, ornamental street plantings, and vegetable and fruit production.

The project was first introduced to representatives from seventy existing gardens on the city's floodplains. Design workshops then defined program guidelines for urban agriculture spaces, gardens in public areas, and gardens for specific soil and land conditions. They discussed land use issues and area requirements—desired uses were clustered together and organized spatially by taking into account their proximity to existing neighborhoods and streets. Large-scale models were created to help residents visualize ideas.

For La Lagunita, six workshops resulted in the development of an educational-productive square that incorporated a playground, sport fields, a meeting and barbeque area, a demonstrative garden, and an educational footpath. The workshop series for the Molino Blanco Sur neighborhood responded to different conditions: this neighborhood already practiced urban agriculture as an income generator, yet the residents had different challenges,

including the periodic flooding of their district. The result of their workshops was to create a new garden-park on the floodplains that used irrigation strategies to protect and water the productive landscape. It also incorporates a soccer field and playground, new community gardens, and walking paths. Lighting, fencing, and a watchtower were installed for safety and security. A greenhouse and market area enable commercial production and sales. Fruit trees have also been planted. The garden-park concept was subsequently applied to 17 hectares (42 acres) of additional city land.

Neighborhood participants were responsible for constructing, planting, and maintaining their new garden/multipurpose spaces. The municipality also helped by passing ordinances that allowed land conversion, continuing training initiatives for the development of organic urban agriculture skills, and slating funding for infrastructure, construction, and maintenance. The design team played a crucial role in facilitating the participation of different stakeholders in conceiving and implementing these neighborhoods as "edible landscapes."

The success of the participatory process in Rosario is evident from new spin-off initiatives being developed using this project as an example: the construction of additional garden-parks along Rosario's ring road, as well as the construction of new garden-parks in Peru and Brazil. From individual houses to whole cities, ideas for urban agriculture and collaborative initiatives can be applied to a variety of conditions.

GREENHOUSE VILLAGE
INNOVATION NETWORK
THE NETHERLANDS

LEFT: Greenhouses designed for the Floriade international horticulture exhibition in The Netherlands, developed around the concept of "building with green and light."

In The Netherlands, the business of cultivating flowers and plants in greenhouses is well established. Greenhouses can collect significant amounts of solar energy, but in the summer much of this is ventilated away to prevent overheating, while in winter many greenhouses are heated, using large amounts of energy to maintain suitable temperatures for growing. This has a significant environmental impact: greenhouses account for almost 10 percent of natural gas consumption in The Netherlands.

Greenhouse Village is a proposal from the Innovation Network in The Netherlands that explores emerging greenhouse technology to develop alternative methods of supplying a community's basic needs based on ecological principles.[50] It considers cyclical approaches and feedbacks in which wastes from greenhouse processes become resources for other needs. The community can benefit from a variety of synergies and closed-loop systems that integrate food, energy, and water supply systems, thereby reducing or eliminating waste. The concept demonstrates how urban environments and infrastructure can be closely connected with agricultural activities.

The Greenhouse Village concept is based on an innovative greenhouse structure that stores excess heat from solar radiation during summer in underground, natural water reservoirs, or aquifers. The greenhouse provides energy and food for a self-sufficient neighborhood and closes water and nutrient cycles at a decentralized scale. The collected heat is used for warming the greenhouse at night or during the winter, but calculations suggest that excess energy would also be available to heat a significant number of homes. A 2-hectare (5-acre) greenhouse may provide heat for up to 200 residences. In addition, the greenhouse supplies tap water, treats

wastewater, produces electricity, and provides employment for residents. It is expected that the whole complex would be self-sufficient in terms of energy—all energy would be derived from renewable sources, such as solar cells and biomass—and water use, and would also recycle nutrients and carbon. Rainwater collection makes up any additional water needs. Wastewater and green wastes are locally treated and reused.

The Greenhouse Village concept is designed around four "Cycles:" Carbon, Energy, Water, and Nutrient. In the Carbon Cycle, biogas from organic wastes is used for heating and electricity generation. Carbon dioxide is used to improve plant growth, and solid wastes are composted and used in the greenhouse. The Energy Cycle harvests excess summer heat for use in winter from the below-ground aquifers. Patented heat exchangers can increase the groundwater temperature from 11 to 27 degrees Celsius while maintaining the air

The Energy Cycle

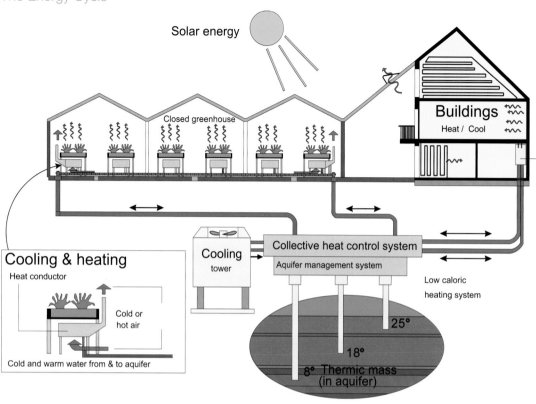

Solar energy

Closed greenhouse

Buildings
Heat / Cool

Cooling & heating

Heat conductor

Cold or hot air

Cold and warm water from & to aquifer

Cooling
tower

Collective heat control system

Aquifer management system

Low caloric
heating system

25°

18°

8° Thermic mass
(in aquifer)

LEFT, ABOVE: Solar
energy harnessed and
stored in underground
aquifers during the
summer months is used
to heat greenhouses
and buildings during
cooler months.
LEFT, BELOW:
Wastewater from
the house is treated
and harnessed to
use for irrigation and
fertilization.

The Water Cycle

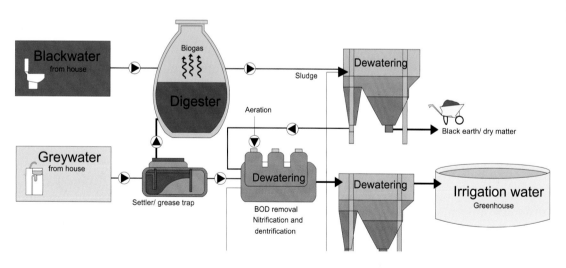

Blackwater
from house

Biogas

Digester

Sludge

Dewatering

Aeration

Black earth/ dry matter

Greywater
from house

Settler/ grease trap

Dewatering

BOD removal
Nitrification and
dentrification

Dewatering

Irrigation water
Greenhouse

temperature of the greenhouse at a maximum of 30 degrees Celsius. The Water Cycle collects graywater from household showers and kitchens for purification in an aerobic bioreactor. Black water from toilets is used to produce biogas that is used for energy. In the greenhouse, evaporated water from plants can be collected, filtered, and used as tap water. The Nutrient Cycle uses the nutrient-rich liquid from the digester to feed the plants in the greenhouse.

Cost calculations suggest that the Greenhouse Village concept may be viable for groups of 200 to 20,000 houses. However, such communities need to be planned so that the greenhouses are located in close proximity to housing and other components in order to reduce losses from heat distribution and pumping costs for fluids. Close integration of greenhouses with the other buildings also offers opportunities for new architectural expression and urban design strategies. In urban locations, city blocks could integrate greenhouses on roof spaces or within tower blocks, combining the various closed-loop systems within a single large building. In low-rise or suburban locations, a lower-density horizontal format may be more appropriate, with housing units close to or attached to a communal greenhouse space. Appropriate passive solar design strategies would be used, and all the technologies would be carefully integrated. Connection with local ecological systems would be a central focus as well.

The concepts behind the Greenhouse Village have also been used by architect Jon Kristinsson to propose an integrated building for the Floriade international horticulture exposition in Venlo, The Netherlands, in 2012. This structure uses new technologies, such as breathing windows and fine-wire heat exchangers from horticulture, to harvest heat from a greenhouse that would supply a four-story office space that forms part of the 15,000-square-meter building. The concept has been adapted to the built environment and is now called Building with Green and Light,[51] making the advantages of combining plants with sustainable energy techniques tangible.

TOP: Innovative housing units are fully integrated into the Greenhouse Village concept and would include features such as living walls.
ABOVE: Amenities such as swimming pools could be provided by the Greenhouse Village's closed-loop systems, as a rendering of this "Sungalow" shows.

SOUTHEAST FALSE CREEK

CITY OF VANCOUVER, HB LANARC, PWL PARTNERSHIP, AND DURANTE KREUK
VANCOUVER, BRITISH COLUMBIA, CANADA

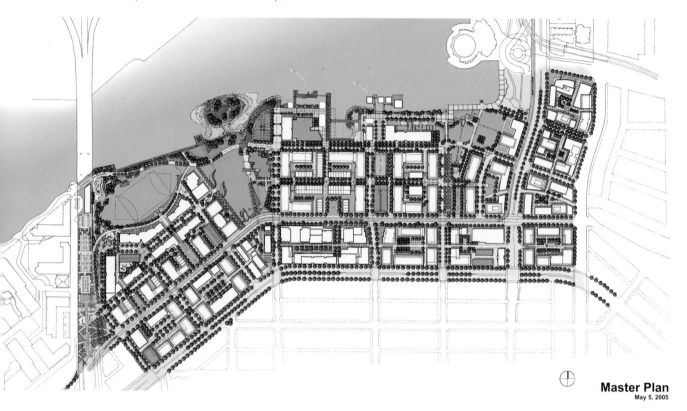

⊕ **Master Plan**
May 5, 2005

Southeast False Creek (SEFC) demonstrates how planning at the neighborhood scale can facilitate the introduction of agriculture into high-density, high-rise urban areas. The area slated to become SEFC is one of the last remaining brownfield waterfront development sites in Vancouver, British Columbia. When complete the 80-acre, mixed-use development will include residential, live/work, retail, offices, public amenities, and cultural venues, and will accommodate an estimated 16,000 people.

SEFC is being developed as a model sustainable community and has achieved Platinum certification under the U.S. Green Building Council's LEED for Neighborhood Development rating system. To help guide the development process, the City of Vancouver commissioned several reports to consider how energy, water, waste, transportation, and urban agriculture will impact the sustainability of the neighborhood. Food issues were considered at an early stage, and, in 1997, the *SEFC Policy Statement* established the central role of food

production in future development of the neighborhood.[52] The objectives were "to establish clarity on the role food production should play in the development of a sustainable city and neighborhood" and "to use urban agriculture and community gardens to assist in meeting other social, environmental and economic objectives."

An urban agriculture strategy was commissioned for SEFC in 2002 to explore opportunities for production, processing, and distribution of food. This document identified spaces for potential food production, including

private residences, public buildings, schools, parks, and street rights-of-way. In 2006 the SEFC Official Development Plan formalized this dedication to urban agriculture, ensuring that opportunities for food production in public and private spaces would be possible.[53]

Holland Barrs Planning Group, now HB Lanarc, prepared a report for the city in 2007 that identifies "viable UA (urban agriculture) opportunities to enhance the sustainability strategies of high-density developments."[54] This report proposes design guidelines, technical solutions and management strategies, highlights the perspectives and needs of the designer and the gardener, and categorizes a range of opportunities for urban food production in private spaces, such as rooftops, balconies, building perimeters, courtyards, and building interiors.

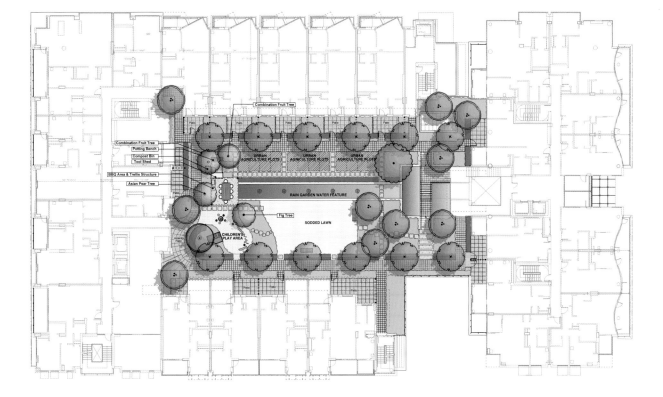

Public spaces, such as plazas, parks, streets, schools, and community centers, are also included.

The SEFC design principles for urban agriculture provide a framework and context to guide more detailed design strategies. The goals are to create an infrastructure that celebrates food production, preparation, and consumption and leads to diverse, productive landscapes. Design should be based on the natural characteristics of the site, starting with the movements of the sun, availability of water, and soil quality. The aim is to improve the ecological health of the site and enhance plant and animal diversity while providing flexible and adaptable spaces that can adjust to future community needs. This design is considered from the perspective of the gardener or user, with attention to spacing, planter design, and access.

The approach taken at SEFC is opportunistic. Urban agriculture is a marketable draw that attracts tenants to the neighborhood. Multiuse spaces keep food cultivation from displacing other activities, while backyard integration enhances social interaction and schoolyards become venues for learning about food. The intent is for urban agriculture to be a visible part of the complex, so any locations where food is grown are designed to be tidy and attractive, and showcase a wide range of food-related opportunities. To this end, technical considerations, such as structural implications for planters connected to buildings and green roofs (and the assurance that only nontoxic materials are placed next to growing areas), the need for irrigation systems, and equipment storage, were also taken into consideration.

The Holland Barrs report also addresses solar and wind exposure, water availability, use of nontoxic materials, accessibility for people, produce, and equipment, and the need for structural considerations and appropriate infrastructure.

The first phase to be completed, known as Millennium Water, covers 18 acres and incorporates the 2010 Vancouver Olympic Village, including 1,100 residential units, 68,000 square feet (6,300 square meters) of commercial space, and a civic center. Although due to economic considerations not all the intended features were included in the final project, the goal of keeping 50 percent of the site green led to 30 percent of the units with a balcony or patio of less than 100 square feet (10 square meters) being given access to at least 24 square feet (2.4 square meters) of gardening space. At street level gardens are also publicly accessible, and in some areas communal crops are proposed. As a result, each land parcel has at least 1,000–1,500 square feet (100-150 square meters) of space for urban agriculture; gardening equipment is provided. A community demonstration garden is also proposed at Hinge Park for school and community programming about urban agriculture. Due to former industrial contamination, a membrane or raised beds will separate the garden from the soil beneath.

In response to public pressure, the City of Vancouver created a taskforce to develop recommendations for urban agriculture throughout the city. SEFC demonstrates the potential of municipal policy to initiate urban agriculture, illustrating how neighborhood planning can stimulate new design ideas that accommodate food growth in dense urban areas. It highlights the importance of a considered approach, goal-setting, and embedding urban agriculture principles into city by-laws and codes.

ABOVE: Courtyard plan for one of the Millennium Water buildings used for the Vancouver 2010 Olympic village. OPPOSITE: The neighborhood's site plan, designed to accommodate food production opportunities.

PARC DOWNSVIEW PARK

PDP INC., MAU DESIGN, INSIDE OUTSIDE, OMA, OLESON WORLAND ARCHITECTS, AND JANE HUTTON
TORONTO, ONTARIO, CANADA

Parc Downsview Park, Inc. (PDP) is a vast, federally owned, formerly industrial property in the midst of Toronto's inner suburbs that is undergoing long-term redevelopment. PDP is meant to serve as a prototype and educational resource for how all of Canada could combine large green spaces with transit-focused, mixed-use sustainable neighborhoods. It is intended as a place for exploration, innovation, and learning for grass roots groups, neighborhood organizations, educational institutions, businesses, and citizens. It also operates as a public corporation that self-finances its development activities. Urban agriculture has been given a significant role in its programming.

PDP sits on land that served for several decades as a hub for civil and military aeronautic research, testing, and construction. While large parts of it are still active, the bulk of the property was decommissioned in the 1990s. Established in 1999, PDP has since engaged with stakeholders, experts, and the public to develop a series of plans for the site. The vision for the 231.5-hectare (572-acre) PDP emerged based on the competition-winning "Tree City" concept for a multipurpose recreational and residential site, which would maintain most of the existing open spaces and reuse the historic, aviation-related buildings.

The goal of integrating urban agriculture into the five neighborhoods that are slated to comprise Downsview Park's built-up areas was only briefly mentioned in its design guidelines. Food production was defined more explicitly as a specific element of its later master plan, when twenty acres were designated as a "Cultivation Campus"—a space meant for both food production and education. This zone would include gardens, greenhouses, a horticultural center, space for a wide range of educational programs, and offices for a variety of community organizations and specialized commercial producers. Local urban agriculture advocates successfully led the PDP Board to strengthen its commitment to the development of food-focused activities. A resulting pilot project on a small site of about three acres, which includes planting in the field and in three new greenhouses, will prepare the soil and infrastructure for the Cultivation Campus in its larger permanent location.

ABOVE: Parc Downsview Park's site plan, showing development of underused land at the perimeter of a former aeronautics site. The Cultivation Campus is slated for the southwestern corner of the site.

OPPOSITE: Rendering of *Hedgerow Project*, a proposed sculptural installation by Blair Robins and Ian Lazarus that would support plant growth, mitigate winds across the site, and create a visual buffer between PDP and nearby roadways.

In addition to classic community and allotment gardens, other types of production areas are planned, including: mosaic gardens, designed as a series of connected pavilions with greenhouses at their core to provide an oasis in the winter months and act as lanterns in the evenings; learning landscapes, which will offer school groups opportunities for hands-on experience working with the earth; and a nursery to serve as a retail model that will emphasize environmentally responsible cultivation and sustainability, innovative growing techniques such as rainwater collection, composting, native plant use, and alternatives to pesticides.

Local urban agriculture advocates successfully led the PDP Board to strengthen its commitment to the development of food-focused activities. A resulting pilot project launched in 2009 by PDP on a small site of about three acres, which includes planting in the field and in three new greenhouses, while the soil and infrastructure for the Cultivation Campus in its larger permanent location are being prepared.

FoodCycles, a nonprofit organization that aims to develop year-round food production and midscale composting in Toronto, was the first group to sign a lease for an acre of land and a greenhouse. Its members and volunteers produce food and compost for donation and sale on-site and through community-supported agriculture. The design, by landscape architect Jane Hutton, features field rows oriented to true north to minimize taller plants shading shorter ones, circulation paths arranged to facilitate the harvesting of produce, and standardized 20-by-100-foot plot sizes to allow for easy seed-purchasing calculations and yields. The main public area—featuring a market stand, workshop space, and eating area—is in the middle of the fields. Paths allow visitors to wander through the fields and the various experimental agriculture and compost demonstration areas without damaging crops. Two entrances from the sidewalk running along a major street at the edge of the site encourage visits.

The greenhouse fulfills a number of support functions, including worm composting and aquaculture.

The pilot project has grown steadily. Urban Harvest, a small commercial producer of organic seeds and seedlings for the urban market, now leases a greenhouse as well. The Ontario Beekeepers' Association has set out twenty-five beehives. Employees of the Toronto and Region Conservation Authority also run a small workplace garden there. PDP itself has been growing trees from locally gathered seed to be used in its own park.

Starting with the winning Tree City competition proposal,[55] PDP has served over the years as a site for experimentation in architecture and urbanism for many artists, architects, landscape architects, and others studying alternative forms of built and nonbuilt environments, which have often included reflections on the place of food production in new forms of urbanism. Recently, PDP has served repeatedly as an experimental field for urban agriculture design proposals put forth by Ryerson University's Department of Architectural Science.

PDP has attracted proposals for art that supports urban agriculture initiatives. *Hedgerow Project*, by local artists Blair Robins and Ian Lazarus, is based on a series of permeable aluminum "curtains" arrayed as both sculptural installation and supports for plants growing in the field. These substantial structures mitigate the harsh winds sweeping into the park and provide a visual buffer from nearby automobile traffic while creating relationships between plant shapes, colors, and textures.

The Cultivation Campus provides much-needed, accessible facilities for urban agriculture at the practical, experimental, and educational levels. PDP is working for recognition as a hub of urban agriculture for locally grown produce throughout the Greater Toronto Area. The implementation of urban agriculture activities has remained somewhat uneven in PDP, yet this project offers an interesting illustration of how a formerly underutilized site can serve as a space for long-term urban agriculture experimentation and education.

ABOVE: Crops are grown in greenhouses as well as open fields.
ABOVE, LEFT: Site plan for the pilot program, depicting greenhouses and field crops, the market and classroom, and paths for community access.
LEFT: Rendering of a student proposal by Megan Albinger showing a recycled fuselage from a Boeing 747 as the structure for a community greenhouse.
OPPOSITE, LEFT: A watercolor rendering of a conceptual greenhouse design.
OPPOSITE, RIGHT: A conceptual, artistic rendering of a vision for allotment gardens.

PRAIRIE CROSSING

PRAIRIE CROSSINGS HOLDINGS CORPORATION WITH GEORGE AND VICKY RANNEY
GRAYSLAKE, ILLINOIS

Prairie Crossing is a 667-acre farming subdivision in Grayslake, Illinois, a suburb forty miles north of Chicago.[56] The project was initiated in 1987 by a group of neighbors who formed Prairie Holdings Corporation to be able to purchase the plot of land due to concerns over an active proposal to erect 1,600 homes on the site. The resulting community of 359 clustered, single-family homes and 36 condominiums was designed to preserve open space, build social capital, and provide public transportation opportunities for a diverse group of residents.

About 400 acres—60 percent of the land—are designated as open space, and include ecologically restored wetlands and prairie grasslands; 125 acres are designated for agricultural use. The original landscape, damaged by years of monoculture farming, was restored by reestablishing hydrological flows, reintroducing wetlands, and introducing organic farming methods between clusters of development. As a result, the ecological diversity of the site has flourished—over 128 bird species and 150 indigenous plant species are now present, compared to the ten bird species and two agricultural crops present before development.[57] The off-site impact of Prairie Crossing had similar benefits: storm water runoff was reduced by 40 percent.

The agricultural component of the project was conceived as part of a larger agenda to develop a model for healthy communities. Ten guiding principles aimed at protecting and enhancing the natural environment and safeguarding open space while providing economic viability and a healthy lifestyle were proposed to balance the needs and health of the residents and the natural landscape. These included opportunities for outdoor activities, indoor exercise, and access to healthy food. The intention is to engage a diverse cross-section of residents in the conservation activities of the community, and to provide resources for learning and personal development. Access to public transport and locally grown food are regarded as key components to creating a community with low environmental impact. The landscape and architecture are inspired by the historic prairies, marshes, and farms of the area.

Prairie Crossing's buildings meet high construction and environmental standards and consume approximately 50 percent less energy than typical new homes in the area. Prairie Crossing is the first community-scale demonstration project to participate in the U.S. Department of Energy's Building America

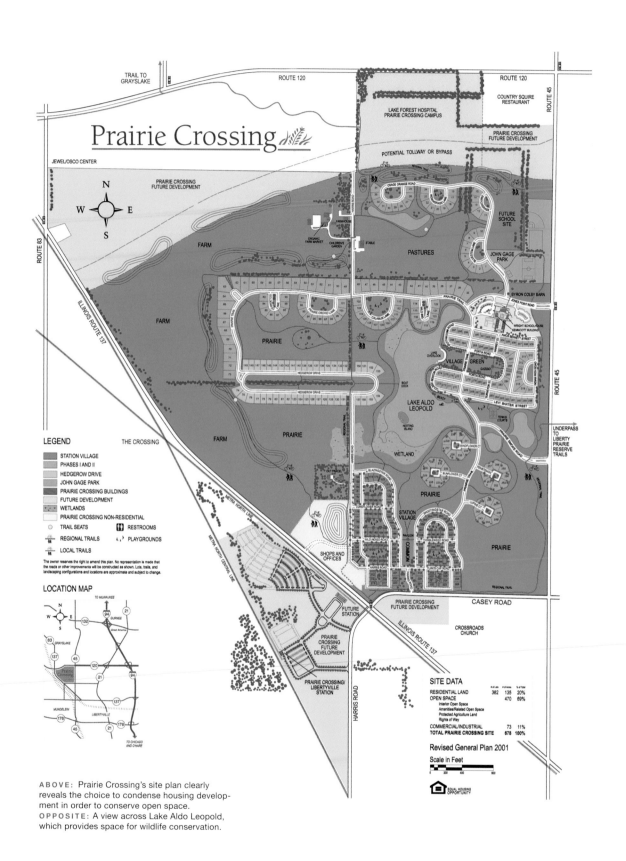

ABOVE: Prairie Crossing's site plan clearly reveals the choice to condense housing development in order to conserve open space.
OPPOSITE: A view across Lake Aldo Leopold, which provides space for wildlife conservation.

program. The housing consists predominantly of low-density, single-family homes that average about 2,700 square feet (250 square meters). The overall density is over 1.5 acres per home with an average lot size of less than a quarter of an acre, but the buildings are clustered to leave large open areas. The architecture was inspired by rural traditions and includes deep porches, clapboard siding, sash windows, and rustic colors with contrasting trim. Recognizing the need for diversity, the community added 113 additional homes in a more dense configuration—although still with 5,000-square-foot lots—and thirty-six condominiums near the commuter train station. Both were designed using New Urbanist principles.

Prairie Holdings Corporation is strongly committed to the cultivation of organic food production. To protect the farmland in perpetuity, a permanent conservation easement has been placed on the land, held by the Conservation Fund, which requires it to remain undeveloped. Appropriate management of the farmland is crucial to the success of farming subdivisions. In the early years, the on-site farm was run by hired staff with volunteer help from the community. In 2004, 40 of the 125 acres of farmland were given in a long-term lease to Sandhill Organics, a family-run organic farm that grows vegetables and flowers and resells fruit from other growers. Although farming in urban areas can create some challenges, such as a lack of support infrastructure and potential friction with residents, it also has many advantages: Sandhill Organics sells about two-thirds of its produce through a community-supported agriculture share system and the rest at local farmers' markets. This reduces

costs for distribution and marketing; it also means that more of the profits from the crops go directly to the farmers themselves rather than to distributors or supermarkets.[58]

The site also includes a 5-acre "Learning Farm" that provides hands-on experience for children from two local schools—agricultural, environmental, and community issues are covered in the curriculum. Sixty acres of land are also used as a farm business incubator that, with support from the Farm Beginnings program developed by the Minnesota-based Land Stewardship Project, provides a management training program for new farmers that helps them create business plans and learn farming techniques from more experienced mentors.

Prairie Crossing demonstrates that locally focused agriculture and residential development can be mutually beneficial. The organization has succeeded in creating a financially profitable model for farmers who are willing to serve local needs, which in turn helps to build a more localized economic base by recirculating local income within the community. Residents benefit from access to healthy local food and reduced food miles. This makes them far more aware of how food systems work, which in turn facilitates informed decision-making about nutritional choices. Prairie Crossing addresses the suburban, or exurban, context for food production. It goes beyond similar "farming subdivision" developments, however, by considering agriculture as part of a holistic approach to reducing the ecological footprint of new housing developments. Agriculture is integrated with energy efficiency, public transport, community support, and healthy lifestyles in a suburban location.

ABOVE: Most housing at Prairie Crossing is low-density, single-family housing.
LEFT: A second phase of development included more density and retail space.
OPPOSITE: Students planting crops at the Prairie Crossing Learning Farm.

AGROPARKS
INNOVATION NETWORK
THE NETHERLANDS

RURAAL PARK

The Netherlands is a small country with a high-density population and a shortage of land. Nevertheless, it is one of the largest exporters of agricultural products in the world due to intensive farming methods that rely on industrial systems. Its small size and the effects of modern industrial farming methods, such as the use of fertilizers, energy consumption, and emissions from farming and processing, have led to concerns about the sustainability of its food supply. Normal organic farming methods make extensive claims on scarce land, and this has led to a growing interest in alternative ways of producing food and integrating food production with other activities.

Innovations in food supply involve not only new technologies but also new administrative, organizational, and cultural concepts. One recent initiative is the "Agropark" concept, developed by the Innovation Network. It is an attempt to develop closed-loop food supply systems at the industrial scale based on intensive food production in limited spaces that integrate agricultural and other economic activities. The approach taken in Agroparks is highly industrialized and controversial, and has raised considerable discussion and resistance. Some aspects of the concept, however, are simply revisiting approaches of

traditional, preindustrial, mixed-activity farms that produced a variety of foods to support their community and were prudent with the use of raw materials and the production of waste. Most of what was produced on a traditional farm found some use as produce for consumption or sale—little was wasted.

Agroparks cluster agricultural functions, including plant and animal production and processing, with other activities to create closed-loop systems that conserve resources and waste flows. Integrated production and food processing at the same location would benefit the environment, people, and animals.

Agroparks are innovative because they link various activities at the same, usually urban, location. They aim to integrate vertical and horizontal activities along the food chain, and partner with other industries, like energy companies and waste management services. Thus farmers, processors, packagers, and even retailers are located together, leading to process efficiencies and decreased transportation costs and animal discomfort, while providing fresh, local food. At the same time, wastes from one production process are used as a resource for another, and symbiotic links are established. For example, animal manure and other organic waste can be digested into biogas that is burned in a power generator, providing heat and electricity for the processing of food, and carbon dioxide could be diverted to greenhouses to increase productivity.

The Innovation Network has been exploring the spatial characteristics of Agroparks in urban and rural environments,[59] proposing four complexes,

from multistory buildings in urban areas to green industrial estates or multifunctional parks in more rural locations. Its Delta Park concept focuses on integrated food production in an urban industrial estate setting to take advantage of available transport, energy supplies, and local markets for fresh produce. Due to high land costs, it employs non-land-reliant agricultural processes in high-density buildings relying on high-tech, IT, and biotechnology solutions. A second proposal, the Agri-Specialty Park, is centered on a sugar refinery located in a harbor area within an agricultural hinterland. Alcohol production or a bio-refinery could occupy space in the complex, both of which process sugar and other products, such as potato, wheat, hemp, and soya, into a range of food and nonfood specialty products, such as pharmaceuticals, cosmetics, and biodegradable packaging. A third concept, the Green Park, is proposed for semirural settings and includes land-reliant animal and plant production and concentrated industrial processing. Its physical form may be more diverse and spread out due to lower land values and densities. Finally, the Rural Park would be located close to, but not in, an urban area. Pigs, poultry, and cows would be integrated with growing flowers, vegetables, and possibly forestry activities or energy production from biomass. The Rural Park would also allow the general public to experience various aspects of the food chain via markets, tastings, exhibitions, and catering facilities. Industrial ecology principles are applied to all the Agropark concepts through the exchange of organic wastes, methane, carbon dioxide, and heat between the various food sectors.

This quote portrays the industrialized and urbanized vision Innovation Network holds for the future and highlights a very particular and controversial interpretation of future food supplies, which contrasts with visions of a more natural, organic approach to farming:

> Former pig farmer Jan Simonse now 'farms' in Amsterdam's dock lands. Five days a week, he commutes by car between home and work. Today, as he nears the harbor he sees a large ship, which has just taken on a huge cargo of cut flowers. Those flowers were grown in the glass houses of a large agropark complex. The glass houses stand atop a building, which houses one hundred thousand pigs. The pigs are free to roam around their pens, each of which has just thirty animals. They can root about in the straw and can enjoy the daylight, which streams in from above. Simonse is one of the managers of this pig farm. He is responsible for purchasing the pig feed, some of which consists of waste and by-products from a nearby food processing plant . . . All the pigs were born here, and all will be slaughtered on the premises. Transport, one of the major causes of animal suffering, is therefore also unnecessary . . . Along with the waste from the glass houses, the manure produced by the pigs is fermented for biogas, which is then used to heat the fish-farming tanks in the basement. Fish of all species and sizes swim here. The basement is also used to grow mushrooms, with pig manure used as the growing medium. The mushrooms produce carbon dioxide, which is then piped upstairs into the glass houses to promote the growth of the flowers.

The Agropark concept generates considerable discussion for its highly industrialized character and concerns about the way animals may be treated and housed in such projects. However, from a design perspective, it could have a significant impact on urban spaces and buildings. New building types that accommodate clusters of livestock, growing spaces, recycling, packaging, and distribution will need to be developed as world population swells, and previously neglected spaces, such as roofs and underused industrial areas, will become valuable for the integration of Agroparks into cities.

PIG CITY
MVRDV ARCHITECTS
THE NETHERLANDS

Pig City is a playful but provocative exploration by Dutch architectural firm MVRDV into how a small country such as The Netherlands may be able to provide healthy, organic food in a world where population is rising and industrial farming methods are being questioned. The project considers the potential for high-rise farming for the particular context of Dutch pork production.

World pork consumption has topped 80 million tons per year, making it the most consumed form of meat. The Netherlands houses about 15 million pigs, a roughly equal number of humans, and is the chief exporter of pork within the European Union, producing about 16.5 million tons per year. Recent concerns about swine flu, foot-and-mouth disease, and the intensive nature of pig farming, as well as growing dissatisfaction among rural residents about the pervasive odor emitted by pig farms and the negative impact of ammonium deposits on green and agricultural areas, have raised questions about meat production and consumption in general. MVRDV estimates that each pig currently requires on average 7,100 square

feet (664 square meters) of land, mainly for intensive feed grain production and industrial processing space. Increasing demand for organically raised pigs pushes the requirements to 130 percent more land, or 18,500 square feet (1,726 square meters) per pig. Fully organic pork production would actually require 75 percent of all the land in The Netherlands to be dedicated to pig farming.[60]

MVRDV suggests that either consumption patterns must reduce meat demand, or production methods must change to find better ways of rearing pigs. The Pig City project speculates about whether it is possible to combine organic farming with densification of pig-rearing activities within "pig apartments"

of concentrated production. By raising production above ground level in high-rise buildings, space is made available below for other uses. Such farms could be located close to centers of pork demand and benefit from the economies of scale by combining facilities such as the slaughterhouses and fertilizer recycling plants.

Pig City differs from other vertical farm projects in that it proposes a discrete high-rise city for pigs, in contrast to projects like the Vertical Farm projects at Columbia University, which are generally integrated into the urban fabric. The Pig City proposal is for forty-four towers, each over 2,000 feet (600 meters) high, with floors measuring 300 by 300 feet (87 by 87 meters). The pigs rummage on large balconies and under trees in open air. A central service core with an elevator allows the pigs to be moved to the abattoir in the base of the building, and a fish farm sited on the roof provides some

TOP: Balconies jutting out from the side of the proposed towers hold several pigs each.
ABOVE: A view of several pig towers from a distance.
OPPOSITE: Pig habitats and maintenance mechanisms designed to make Pig City a carbon-neutral, closed-loop system.

of the food. Pig manure is collected to a central slurry-processing plant and used in a biodigester to provide the tower's energy, reducing odor and ammonium pollution.

The researchers estimate that each tower could feed up to half a million people, so thirty-one Pig Towers would supply the entire Dutch population with meat. The remaining towers would produce enough pork to maintain present export levels. To minimize transportation requirements, many of the towers would be positioned near the Rotterdam and Amsterdam harbors or on land reclamation areas designated for port-related industrial zones. Other towers could be distributed throughout the rest of The Netherlands. The towers, including the surrounding reed fields that process the waste generated, would take up just over 5 percent of Dutch land area.

Pig City is a radical and utopian concept that demonstrates the potential of high-rise buildings to address problems of density and space constraints while bringing to the fore many of the same concerns about the quality of life for inhabitants of residential towers that have dogged them for years. Some have questioned the entire concept of industrial "food factories" and the desirability of housing animals high above ground in exposed conditions. Animal welfare was an important factor in the development of this concept, however. The designers argue that conditions here are better for the animals than at farms: providing each animal more space and avoiding the transport of live animals are both central issues in animal welfare discussions. Rural areas would also benefit from reduced odor and harmful emissions and ammonium deposits.

Concerns have also been expressed about creating another monoculture that focuses on only one product—pork. Organic methods are inherently more diverse and rely on mixing a variety of plants and animals. Since feed production uses the most land in agriculture, questions arise as to whether these structures would change much because the pigs must still be fed an equal amount, and there are limits to how much can be produced within the building. Despite the industrial approach embodied by Pig City, natural systems for providing energy and food were indeed also considered. A final issue relates to the environmental cost of creating these buildings. Tall buildings embody much energy in the form of construction, and they require a lot of industrial materials. The impact of these also needs to be weighed against the potential benefits to the food chain.

Pig City is part of a growing group of provocative, high-rise farm projects that have been proposed around the world and that attempt to embrace organic methods within a factory farm setting. This concept, from an architectural practice that adopts a critical thinking approach to current world problems, can be interpreted as "a stern warning about what might lie ahead should modernist principles be applied on purely economic grounds and not ethical ones."[61] The concept continues to be explored by members of MVRDV and The Why Factory at Delft University in their Pig City projects.

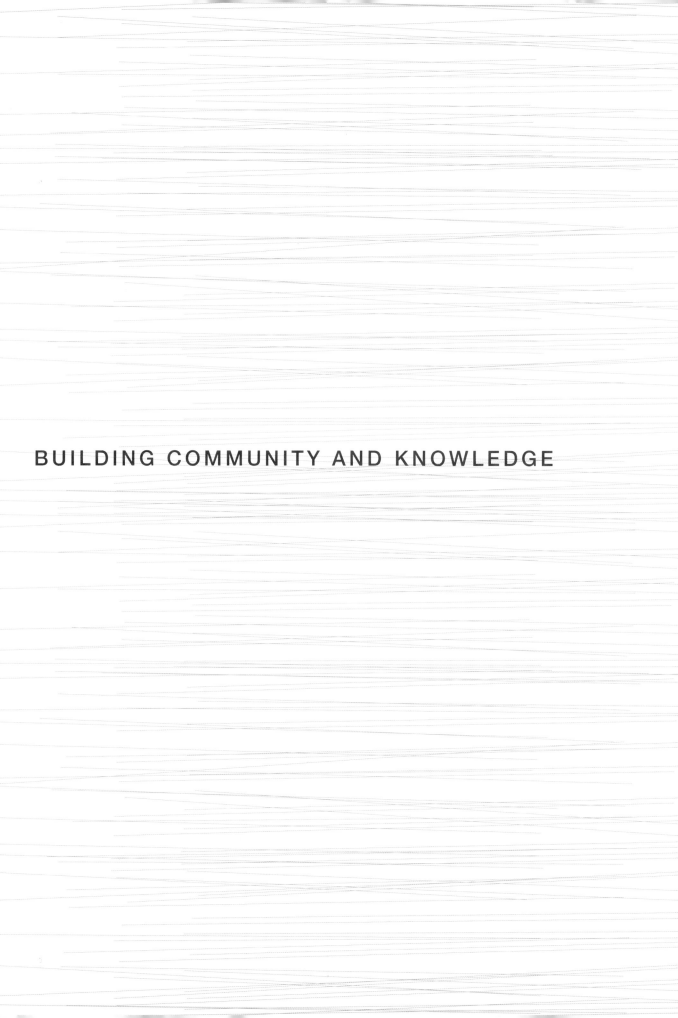

BUILDING COMMUNITY AND KNOWLEDGE

Shared gardens bring neighbors together for a variety of activities that go beyond the act of cultivation. These spaces, whether formal allotment gardens or unorthodox found plots, are increasingly seen as vehicles for forming and strengthening communities, as well as sites for sharing knowledge, learning, and leisure. Many recent examples of gardens and farms, related structures, and organizations that support community through gardening and similar endeavors emphasize the role of design and designers in strengthening the contribution of food production to building community and knowledge in cities.

The act of creating a garden, not merely using one, can also create social bonds. The cooperation of many devoted people saved 31 acres of state-owned surplus land in Madison, Wisconsin, from suburban tract development. The result was Troy Gardens: an integration of mixed-income sustainable housing, 330 family garden plots, herb gardens, a children's garden, an organic farm based on the community-supported agriculture model, and restored prairie and woodlands (see page 64). The architectural design for the housing incorporates common courtyards, and the landscape plan provides attractive walking trails around and through the prairie, woodlands, and gardens to promote social interaction.

Food and community needs are addressed in a surprisingly different way at the Inuvik Community Greenhouse in Canada's Northwest Territories. An

adaptive-reuse design repurposed a former hockey arena—lying just north of the Arctic Circle—into a 4,000-square-foot greenhouse. A commercial operation produces bedding plants and hydroponic vegetables, profits from which cover the costs of raised garden plots that are used by a range of community members, including tribal elders, residents of group homes, and children's groups (see page 67).

Not all gardens that serve the community need to be publicly owned. Gardens run by local restaurants can act, like coffee shops, as social hubs. Roberta's, a pizzeria in Brooklyn, New York, does just that. The restaurant makes a point of buying vegetables from local producers only—including the newly launched Brooklyn Grange, an acre-sized rooftop garden nearby. The owners also grow their own vegetables and herbs in an adjacent garden, and in greenhouses atop shipping containers in the restaurant's

ABOVE: The Plant Room, a What if: projects tomato plant stand placed in a neglected space beneath a billboard in London.
ABOVE, RIGHT: Greenhouses installed over shipping containers at Roberta's Pizza in Brooklyn, New York.

Urban Agriculture Creates Social Bonds

Gardening is an activity proven to create community. London's Plant Room, by What if: projects, is one example. A small, unassuming paved patch of land under a large billboard near a housing project holds a collection of potted tomatoes that sit on basic shelving. Open access and the introduction of a shared barbecue were all the catalyst needed to persuade neighbors to socialize in the space. As Carolyn Steel reminds us in her book *Hungry City*, the word "company" derives from the Latin words *com*, meaning "with," and *pan*, meaning "bread," which suggests that sharing food is a natural way to form bonds.[2]

yard. They also sell produce and bread. The restaurant constantly searches for ways to promote urban agriculture and even hosts Heritage Radio Network, a Web-based, food-oriented radio station in one of the shipping containers. In Brooklyn social connections between gardeners may be almost entirely virtual, but they are strongly supported by this brick-and-mortar restaurant.

Urban Agriculture and Community Involvement

Community gardens contribute to the health of the population they serve in two ways: by yielding fresh, healthy vegetables and by providing a place for neighbors to gather, enjoy nature, and socialize as they learn from each other and develop relationships. They often allow people to participate in programming that includes not only gardening and composting workshops, but shared meals, holiday celebrations, flower festivals, photography competitions, performances, movies, and toddler playgroups.

While the basic aim for gardens is to encourage participants to share knowledge, techniques, land, tools, and water to produce healthy, nourishing, and culturally appropriate food, the social and educational activities associated with food production in shared gardens can be as important as the production itself. Urban agriculture initiatives are often linked with other social equity movements and support small-scale, community-based programs such as start-up businesses or "food incubators" and after-school programs for at-risk children.

The social benefit of community gardens also has been studied scientifically, and their positive impact on neighborhoods documented. Four Canadian researchers, for example, investigated the dozens of gardens scattered around Regent Park, a 1960s social housing project in Toronto that is now

undergoing a radical transformation.[3] The study, called *Growing Urban Health: Community Gardening in South-East Toronto*, highlights numerous benefits that Regent Park's residents experienced from community gardening, including better overall nutrition, improved mental health, a sense of community cohesion, and increased physical activity. Researchers at the University of California's Cooperative Extension also note from their own research that "community gardens beautify neighborhoods and help bring neighbors closer together. They have been proven as tools to reduce neighborhood crime—particularly when vacant, blighted lots are targeted for garden development. Community gardens provide safe, recreational green space in urban areas with little or no park land, and can contribute greatly to keeping urban air clean."[4]

Community gardens generally epitomize the idea of self-design by gardeners—traditionally citizens make individual decisions on the layout of each plot, with community garden leadership intervening only to determine the garden's overall form, which is usually a basic grid. In recent years, groups have started to work with professional designers in order to realize

ABOVE, LEFT: The Harrison Urban Garden in Boston.
ABOVE: Rendering, top, and image of a built pavilion, bottom, that does double duty as a sundial, built by student volunteers at the Worcester Street Community Garden in Boston.

a space that appeals to the entire community, not only active gardeners. They design both decorative and functional elements: places to relax in the shade, tool storage, vegetable washing stations, rainwater collection facilities, and activity spaces.

The South End Lower Roxbury Open Space Land Trust (SELROSLT) is a membership-supported, nonprofit organization that has been working since 1991 to acquire, improve, and maintain open space for community gardening and pocket parks in one of the poorest parts of Boston, which is today being transformed through gentrification and redevelopment. In recent years, SELROSLT has worked with architects and landscape architects on a number of its gardens. In one case, over two summers, it acted as a client for the Boston Architectural College's Urban designBUILD program, where design studios were combined with hands-on construction of small structures. For the first studio, at the Worcester Street garden, a multipurpose pavilion was erected to provide shelter, shade, and storage while providing a gathering space in the heart of the garden. It is also cleverly formed to project shadows like a sundial. The scale and context of support structures that respond to specific needs in community gardens proved perfect for the educational purposes of a design-build course.

In a second case, SELROSLT became the client of an architectural firm and a landscape architecture firm. The group was awarded funding from the Environmental Protection Agency to remediate a contaminated brownfield site that was formerly home to a battery factory, and designed the newest community garden in Boston's South End, the Harrison Urban Garden, completed in 2008. The City of Boston, original owner of the land, selected the developer to create a mixed-use residential/commercial project with the stipulation that the development include a permanent home for an existing SELROSLT community garden that had to be displaced. The resulting garden, containing 32 individual plots plus one common plot, can actually be considered a ground-level green roof project since it rests over an underground parking garage. A 3-foot (90 cm) deep engineered-soil medium replaced the site's original contaminated soil and is placed over drainage mats and waterproofing to protect the garage below. The design drains away standing ground water to reduce load, but therefore also requires diligence on the part of the gardeners to keep beds adequately mulched and well watered. The garden was carefully sited and configured to minimize effect of the shadows cast by surrounding tall buildings so that the plants receive sun for most of the day. Since the community garden is just one part of the larger mixed-use development, in addition to providing plots for growing food, it was also designed to provide public space for residents of the newly revitalized neighborhood.

Another example of community building through professionally designed gardens is the Curtis "50 Cent" Jackson Community Garden in the New York City borough of Queens. Long before landscape architect Walter J. Hood, Jr., created its current design, neighbors had already cleared the neglected vacant lot and tilled the soil to make productive plots on their own. The dramatic renovation, completed in 2008, incorporates functional elements while adding park space to the neighborhood (see page 70).

In Manhattan, the West Side Community Garden, which involved collaboration with artists and landscape architect Terry Schnadelbach, is tucked away midblock on West Eighty-ninth Street. The garden's variety of programming and features include a tulip festival and photography contest every April, children's planting beds, tool sheds, and composting areas. Volunteers keep the garden's common areas tidy and beautiful. Arts and crafts fairs, musical performances, and a children's summer Shakespeare festival with performances in the garden's oval Floral Amphitheater attract residents who might not otherwise have reason to visit the space. Well-designed gardens provide places for the wider community to relax and mingle, and provide much-needed green space in dense urban areas.

Community Food Centers

Community Food Centers (CFCs) are organizations that provide a forum for community participation, engagement, and education about food and agriculture, and provide support networks that can help to create and maintain community gardens. CFCs offer practical support and training for neighborhood residents and contribute to a secure and sustainable future for urban food systems. One of the best-known CFCs is The Stop Community Food Centre in Toronto. Its new home in the Artscape Wychwood Barns (see page 74) supports extensive outreach programming, including gardening, food processing, and community activities. At the Barns, classes help children grow their very first sprouts and help adults learn to make preserves. Local executive chefs also introduce the populace to gourmet cooking techniques through a variety of workshops. The Niagara CFC proposal (see page 84) speculates on the design such facilities could take in the future.

In addition to CFCs, public and private community organizations have begun actively incorporating gardening into their missions. Evergreen Brick Works in Toronto (see page 78), for example, is a reclaimed factory site that uses art, nature, and heritage programming to draw people to the multipurpose facility dedicated to promoting sustainable living. Innovative gardening spaces, a retail garden center, and a children's demonstration garden were carefully planned for the space.

Gardens for Demonstration, Education, and Inspiration

Demonstration gardens support community urban agriculture by providing education and inspiration. One of the most unique demonstration gardens in form, the Science Barge, currently anchored in the Hudson River near Yonkers, New York (see page 86), is a sustainable floating farm with the latest in hydroponic growing techniques and renewable energy systems. The barge produces zero net carbon emissions, zero chemical pesticides, and zero rain runoff. Visitors are shown how to reproduce these low-impact food production techniques in their own communities.

Programs that incorporate productive gardens at schools and universities are in a unique position to spread knowledge about appropriate practices in food production. In North America, many elementary school gardens are modeled after the Edible Schoolyard program initiated in 1995 at Martin Luther King, Jr. Middle School in Berkeley, California. Two years of planning and work by chef Alice Walters, principal Neil Smith, and teachers turned an acre of asphalt into an Eden of fruit trees and vegetables, thanks to the receipt of community and foundation grants. The nutritional value of the school's meals improved, and, equally important, children learned about growing, cooking, and eating healthy meals firsthand. The Edible Schoolyard organization is now in the process of creating a second education-based garden at a school in Queens, New York (see page 90).

Detroit's Catherine Ferguson Academy, a school for teenage mothers, also features a garden as an integral part of its curriculum. Life sciences teacher Paul Weertz created a small farm on the school campus that includes a broad range of farm animals, orchards, vegetable and flower gardens, and beehives. Students learn about caring for the young by monitoring baby animals and learn economics and marketing by selling their crops at market. In Detroit in particular, residents have seized upon gardening as a vital community builder and a way to productively use vacant lots that have resulted from deurbanization. Ashley Atkinson from Detroit's Garden Resource Program Collaborative notes, "I used to have the foolish idea that urban gardening was all about the food. Now I think that food is only a small part of it. Gardening here is about beautification, community building, friendship."[5]

Universities are also ideal sites for interaction between students and community members. The Warrior Demonstration Garden at Wayne State University in Detroit and Canada's examples at McGill University (see page 94) and Université du Québec à Montréal (UQAM) show how vegetable gardens can become ornamental parts of university campuses as well as learning opportunities for students and a practical source of food for campus dining halls and city food banks. In the summer of 2007 at UQAM, a group of environmental studies students took it upon themselves to transform parts of the campus's manicured lawn into planting beds for edible crops—without authorization. The number of participants grew quickly and even included some faculty. This act of guerrilla gardening was initially resisted by the university administration but eventually evolved into a sanctioned means of campus beautification. The group now composts all of the garden waste produced by landscapers across the campus—and indeed the remaining lawn shrinks every year as the initiative expands. University leadership now provides financial support and a summer school on urban agriculture is held there annually.

An example of a garden designed to teach adults can be seen at Growing Home's Wood Street Urban Farm in Chicago. Its not-for-profit mission is to educate the homeless and unemployed by offering participants six months of farming employment while providing both job and organic agriculture training to give them a marketable skill set. A new multipurpose

LEFT, BELOW: Vegetable garden edging by an academic building at Université du Québec à Montréal.
RIGHT, BELOW: Vegetables growing at Hart House at the University of Toronto.

building has helped support the dual mission of farming and education (see page 98).

Working professionals can also enjoy the benefits of community gardens. Mitchell Taylor Workshop proposed a temporary urban farm for London's financial district on a construction site whose future office tower had been stalled by economic woes (see page 100). Although the project was not realized, it highlights simple ideas for changing the nature of a dense "concrete canyon."

High-profile demonstration gardens in civic spaces can also encourage the creation of productive gardens. For several years, Paris hosted a lavish temporary garden in front of the Hôtel de Ville that featured an extensive educational component. In Vermont 2009 budget cuts left little taxpayer funding for the maintenance of the State House lawn, so volunteers planted a vegetable garden in former flower beds, making it the first U.S. state capital to boast an edible landscape.[6]

The most famous recent demonstration garden at a civic institution is the White House kitchen garden. This is not without precedent—John and Abigail Adams and Eleanor Roosevelt both had gardens of note. An early wooden greenhouse, built in 1857, and a later iron-and-glass conservatory had to be dismantled in 1902 to make way for the construction of the West Wing. Michelle Obama's vegetable garden attracts attention because it is an indicator of the widespread reemergence of interest in local organic gardening to save energy and improve nutrition. Since its groundbreaking in March 2009, it has been used as a way to teach local children how to plant, tend, harvest, and cook. Much of the garden produce is donated to local food banks and soup kitchens, raising awareness about food insecurity even among the United States's citizens. In 2010 the first lady expanded the garden, a sign of her commitment to this campaign, as well as its success.

Another highly publicized and inspiring project, Public Farm 1 (PF1), was designed by WORK Architects and was displayed at the PS1 branch of the Museum of Modern Art. Its innovative form and design—which used ordinary construction materials—resulted in a striking, interactive container garden (see page 104). Interventions like this, helped by museums and galleries,[7] have been particularly effective at demonstrating how aesthetically compelling designs can create space for food where hardscaped public space is the only land available.

Conclusion

Four components are known to be crucial to the creation and maintenance of community: spirit, trust, trade, and art. In the context of community gardening, spirit can be defined as a sense of belonging and a feeling of friendship and connectivity with others in the group; trust as a sense of safety, that the group environment is a place to be oneself and a place where the individual is respected; trade as the perception that group members can both give to and receive from the community such things as ideas, support, knowledge, friendship, and other elements; and art as community values and shared experiences that can be expressed visually or through dance, storytelling, or music. In addition, the development of networks is crucial to

ABOVE: A 2008 demonstration garden installed in front of the Hôtel de Ville in Paris to raise awareness about urban agriculture.

generating social capital and maintaining any sense of community. These are key to the dissemination of knowledge and skills, which improve the quality of life for all connected.[8]

Teaching or learning about gardening techniques, attending composting workshops, taking canning or cooking classes, and going to picnics or cookouts strengthen the links between those participating in urban agriculture initiatives. Outreach to potential gardeners via demonstrations and festivals increases ties; current participants relate the pleasures of gardens in the vicinity, even if newcomers are there to simply enjoy the beauty, quiet, or shade. Community gardens and other spaces of social interaction centered around food production are shown to have both social and health benefits for participants. As examples in this chapter show, designers play an important role in building community and knowledge by designing the components that make community gardens pleasant and functional places.

TROY GARDENS

ZIEGLER DESIGN ASSOCIATES, COMMUNITY GROUNDWORKS, AND
THE MADISON AREA COMMUNITY LAND TRUST
MADISON, WISCONSIN

Troy Gardens is a unique 31-acre housing development that integrates mixed-income cohousing, community gardens, an organic community-supported agriculture (CSA) farm, and restored prairie and woodland open space. Located in Madison, Wisconsin, it was born of a community vision for a unique parcel of state-owned surplus property. The architecture and landscape design tie many land uses together in a harmonious way. Shrubs, for example, were carefully placed to provide wildlife a corridor through the site. Interpretive walking trails showcase plants that are suitable for personal perennial gardens and infiltration, or rain, gardens. Two herb gardens were placed near a main drive; they are ornamental as well as practical: residents can drive or cycle by and clip sprigs.

Troy Gardens has a long history of contestation, negotiation, and collaboration. In 1995–96, the state of Wisconsin placed this site on its surplus land list with the intention to sell it for development. Madison residents had already been gardening on four acres of the site for fifteen years and using much of the rest for community purposes. Alarmed at the prospect of losing this valuable resource, the community rallied. Several not-for-profit organizations—including the Madison Area Community Land Trust (MACLT) and the Urban Open Space Foundation (UOSF), now the Center for Resilient Cities—joined together with representatives from the University of Wisconsin-Madison to form the Troy Gardens Coalition.

The coalition developed an innovative proposal for integrated land use, and in the summer of 1998, the coalition and the state reached an agreement for a fifty-year lease, with a provision to buy the property. Later that fall, Madison's Common Council formally adopted the coalition's concept plan for Troy Gardens, cementing the community's vision. Between 1999 and 2001, the UW-Madison Department of Landscape Architecture devoted several design studio classes, including sessions on open space design and restoration ecology, to develop design concepts for the land that were subsequently presented to the community for feedback. In 2001 the Friends of Troy Gardens (FTG), successor to the coalition, was incorporated as a nonprofit organization. In partnership with MACLT and UOSF, FTG developed and managed the open space. Its mission is to develop a world "where all people have the tools to grow food and steward the environment."

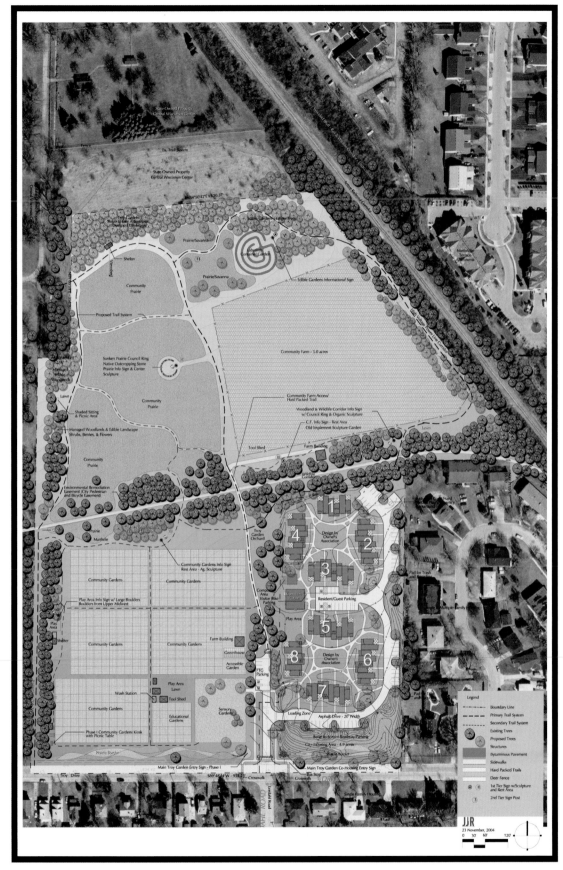

LEFT: The Troy Gardens site plan, showing the large percentage of the original site that has been preserved for community gardening, bottom left, for an organic CSA farm, top right, and as open space, particularly in the form of the restored prairie in the top left.
OPPOSITE, LEFT: Cohousing units fronted by a border of native prairie plant species.
OPPOSITE, RIGHT: A mosaic tile sign announces the entry to the Children's Garden, one of several contributions by local artists.

Immediately upon the purchase of the land, MACLT entered into a conservation easement with UOSF to protect the 26-acre open space portion of the site from development for anything other than parkland or urban agriculture. At the same time, it entered into a ground lease with FTG so that the latter might continue its work to implement the community's vision for the land.[9]

The approach taken to restore the natural areas was unique in that it was designed in partnership with the community—Northside residents, neighbors, and organizations gave input and helped lead the process. Landscape architecture firm Ziegler Design Associates used the results of a final charrette attended by eighty-five participants to complete the design process for the site and to prepare the restoration and management plan for the natural areas.[10]

Neighbors care for 330 family garden plots in the Community Gardens. Volunteer stewards restore and maintain native tall grass prairie and maple woodlands in the parkland areas. In addition, over one hundred households pick up weekly bags of fresh organic vegetables from the 5-acre Troy Community Farm, established in 2001 as Madison's first urban farm. It is operated on the community-supported-agriculture model, through which farm members receive fresh, locally grown, organic produce each week from June through October.

Three additional edible landscapes were installed in 2003 to 2004. The Edible Woodlands is a multilayered system of canopy trees, understory trees, shrubs, and ground cover that mix edible with decorative species for a taste, smell, and visual experience. A path takes residents on a tour of mature sugar and silver maples, as well as other species retained from the initial clearing of the site, including walnut, mulberry, black cherry, hackberry, Russian olive, black raspberry, sumac, and asparagus. Introduced plants are a mix of edible and unique ornamental species, including nut trees and shrubs, fruit trees, berries, and perennials. These were selected for edibility and food production, colonization and woodland development, and their ability to be self-maintaining. Schoolchildren working with a UW-Madison intern also created the Herb Garden. Here, interconnected woodchip paths, winding through a variety of theme beds, meet in a central gathering space. Plants include culinary herbs, traditional medicinal herbs, herbs for teas and drinks, perennial fruits and vegetables, and plants used for textile dyes, health, and body care. A group of Hmong community gardeners tends a more specialized herb garden, with winding paths and retaining walls that divide the space into seven planting areas, creating places for relaxation and contemplation, as well as education and access to traditional Hmong herbs.

Many residents of the thirty cohousing units on the site use the gardens regularly. These residences have environmentally sensitive features and are sited around common courtyards to generate a sense of community. The units are confined to a small section of the site to maximize the area devoted to conservation and food production. MACLT owns the cohousing land and open space. To provide housing for low-income community members, this land is held in a trust that enables the homes to sell for below market rate.

Community GroundWorks, formerly known as FTG, now conducts environmental education programs that include a nationally recognized leadership program for teenagers, an award-winning children's garden, and an innovative partnership with UW-Madison. The Kids' Gardening Program provides gardening, arts, nutrition, cultural, and environmental education to Madison's youth. Children from area community centers plant and maintain their own garden beds and participate in arts and crafts projects on site. Produce from the garden is used for cooking lessons, community meals, and donations to community centers and food pantries.

The community's original vision has taken firm root and has been secured in perpetuity through MACLT's ownership of the site, the Center for Resilient Cities' conservation easement, and the long-term stewardship commitment of Community GroundWorks and the Northside community. As Ziegler Design Associates puts it, "Troy Gardens is many projects rolled into one. It is about feeding a community with a culturally and economically diverse population and teaching residents—both young and old—the skills to grow, prepare, preserve, and sell their own food, and to care about the environmental resources around them. It is about growing community ownership and cultivating a sense of place. It is about community residents and local institutions working together to preserve, sustain, and strengthen their community."

INUVIK COMMUNITY GREENHOUSE
COMMUNITY GARDEN SOCIETY OF INUVIK
INUVIK, NORTHERN TERRITORIES, CANADA

Just above the sixty-eighth parallel, in the Arctic Circle, lies the Inuvik Community Greenhouse, a garden that serves a small town of 3,700. This adaptive reuse garden structure was, at its inception, the northernmost large-scale greenhouse in North America. In 1998 local residents of the rural town of Inuvik proposed creating this greenhouse from a structure that formerly housed a hockey arena. The building was slated for demolition, along with the adjacent structures of the former Grollier Hall residential school. Two proponents of this project, Ron Morrison and Peter Clarkson, are gardeners with greenhouses at home who saw the potential for the arena's transformation. The next year, with public and private support, renovations commenced upon the vast Quonset-hut-shaped building. Clarkson describes the work: "We had a foreman and five workers, and there they were, two degrees above the Arctic Circle in minus 35-degree weather, tearing down boards and metal and insulation—converting a hockey rink into a greenhouse. The sun doesn't even come up at that time of year. People thought we were insane."[11]

One of the most important adaptations was to replace the existing tin roof with polycarbonate glazing. While that material had to be purchased new, the project also reused an estimated $61,000 of building materials from the Grollier Hall structure. Recycled materials were used for floors, windows, and doors. In addition, the recycling initiative extended to the first planters for the bedding plots, which were made from recycled drink cups and milk cartons. Compost for the garden soil is also made continuously from food waste.

Although the growing season in this extreme northern climate lasts only from April to October, during June and July the sun shines for twenty-four hours a day, heating and lighting the space. The plants

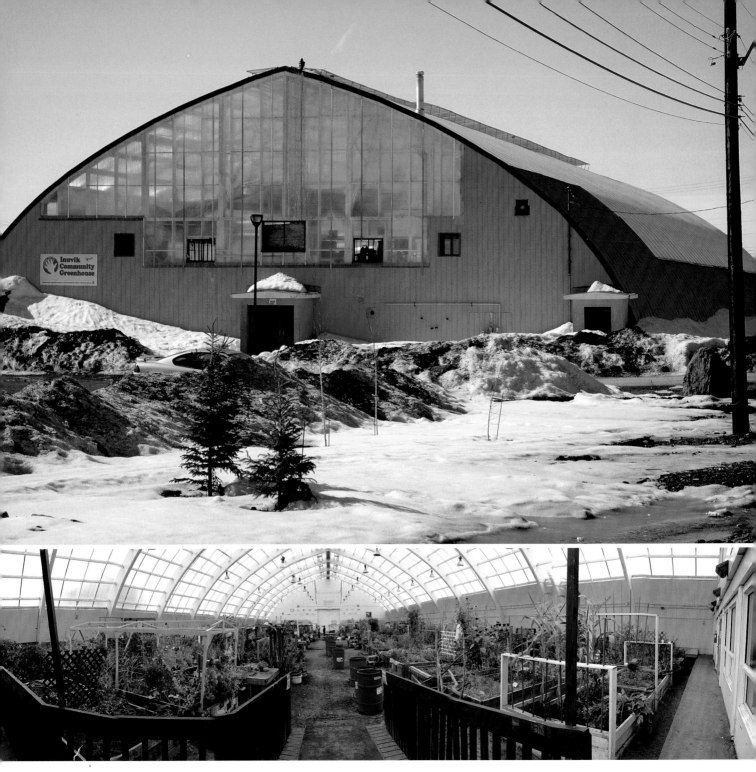

TOP: The exterior of the Quonset-hut-shaped, converted greenhouse. The presence of snow hints at the structure's location north of the Arctic Circle.

ABOVE: A panoramic view of the greenhouse's interior showing the numerous plots in current cultivation.

PREVIOUS PAGE: Blue barrels filled with water act as heat sinks that absorb solar radiation during long summer days and radiate heat back to the plants at night, keeping the interior temperature stable.

thrive under such conditions. Seventy-five raised-bed garden plots sit on insulation placed over the gravel floor above the permafrost, filling the structure. On the second floor, a 4,000-square-foot commercial farm uses hydroponic growing techniques to produce ornamental plants as well as vegetables. This operation adds to the financial viability of the project while also adding to the beauty of the town by selling flowers and potted plants for civic buildings and private businesses. The number of plots in the community garden is insufficient for the size of the population, however, and there is a waiting list. While it is unfortunate that not every resident who wishes access can be granted a plot, it is also a sign of the project's success.

The nonprofit Community Garden Society of Inuvik manages the greenhouse. Their motto, "Promoting Community Through Gardens," succinctly summarizes their vital mission. Citizens can become members of the greenhouse for a small fee and fifteen hours of volunteer time per year toward activities that include making compost, construction, filling water barrels, helping elders or children with their plots, fundraising, hosting greenhouse events, and committee work. A local elementary school also tends a plot, and community groups—including the food bank, the youth center, the high school, and the senior center—raise vegetables in the greenhouse as well.

The society contributes to the sense of community fostered by the greenhouse by organizing many activities. A plant sale opens the season, and, during the summer, a Garden Market is held every Saturday. Tourists visit the space; buy local crafts, vegetables, and baked goods; and can view the display of the local quilting society's work in the greenhouse's classroom area. The society also holds workshops on basic gardening, creating hanging baskets, composting, and nutrition. These activities are organized to add to the knowledge base of the gardeners, of course, but they also bring people together.

Partner organizations help the greenhouse survive. Aurora College was one of the main supporters of the initial project. The Aurora College Trades Access Program continues to provide student labor from the college, while the greenhouse provides materials for the ongoing maintenance of the structure's carpentry, plumbing, and electrical systems. Techniques that help the greenhouse work in extreme weather conditions include an automatic roof vent, installed to cool the building on hot, sunny summer days, and a thrifty but effective passive solar heating system composed of plastic garbage cans full of water that radiate heat during the night.

The success of this greenhouse project can be measured in part by the number of greenhouses that have since sprung up in other northern towns. These include the 90-square-meter, purpose-built Iqaluit, Nunavut Greenhouse, which opened in 2007; Carmacks, Yukon, a small town north of Whitehorse, whose greenhouse provides fresh vegetables to its small community; and a specially reinforced greenhouse at the Devon Island outpost—used for simulating conditions of future Mars expeditions. In this small greenhouse, researchers have been experimenting with innovative growing techniques since 2002.[12]

The Inuvik project demonstrates that with the proper support, coupled with community-based initiatives, creativity, and energy, a town can benefit from a community greenhouse, not only by raising the level of nutrition through access to fresh food, but also through the interactions between community members. The benefits of long-term health and well-being for a community outweigh the work involved in tending the plots and disseminating gardening knowledge. The project also provides a lesson about the benefits of adaptive reuse, which can turn an abandoned facility into a vibrant community hub.

ABOVE: Ornamental plants grown in the greenhouse are sold back to the community to support the greenhouse financially.

CURTIS "50 CENT" JACKSON COMMUNITY GARDEN
WALTER J. HOOD, THE NEW YORK RESTORATION PROJECT, AND THE G-UNITY FOUNDATION
QUEENS, NEW YORK

" Every child should have access to open, green space where they can learn, play, and be active in their community."

CURTIS "50 CENT" JACKSON

ABOVE: The garden, set in the midst of a Queens neighborhood, functions as a hub of community activity.
OPPOSITE, TOP: The garden's site plan.
OPPOSITE, LOWER LEFT: Specially designed cisterns gather rainwater and funnel it to underground holding tanks.
OPPOSITE, LOWER RIGHT: Ornamental plants mixed with edible plants are arranged into visually appealing blocks of color.

In 2007 an established community garden in one of Queens's low-income neighborhoods was thoroughly transformed. Initially known as Baisley Park, it dates to the early 1990s, when neighborhood residents cleared an almost 1-acre plot of wasteland to create a space for productive growing. Donations and the imagination of professional designers, including Walter J. Hood Jr., the Californian landscape architect, helped to create this distinctive garden. Inaugurated in 2008, the redesigned park stands out as an aesthetically compelling community hub for children and adults.

Funding for the project came from the G-Unity Foundation, which was started by rapper Curtis J. Jackson III, better known as "50 Cent," who wanted to give back to the community where he was raised. The joyful design makes this innovative garden more than a place to grow flowers and vegetables: it has also become the community's anchor and showpiece, a place for gardening

workshops, movie nights, and events. It is also, of course, a place where many gardeners meet and connect with nature and each other.

Cheerful, bright colors designed to appeal to the local schoolchildren who would use this as a learning garden and contemporary planting schemes inspired by the Kitchen Gardens of the Château de Villandry in France define this

wedge-shaped garden. Its most distinctive and notably sustainable feature is the cluster of bright-blue, 10-foot-tall rainwater collectors that funnel water to two 1,500-gallon underground cisterns, reducing the reliance on city water to maintain crops. The rainwater collectors are key to the project's success. These huge metal and fiberglass funnels are anchored by reinforced concrete piers spaced between the cisterns and are covered with wire mesh screens to keep out debris. Positioned over a bed of small rocks and pea gravel, the design enables good drainage as well as water retention and provides a shaded play space for children.

Wildflowers/ Annuals Planting

12'X4' Compost Bin

Red Clover

Storage Shed (reused container)

Concrete Pavers

Lawn

Hedges w/ Floral Planting

Gazebo/Rainwater Collection

Gravel

Arbor w/ Fence and Gates

Planting Strip w/ Street Trees

New Sidewalk w/ Curb

LONG ISLAND RAILROAD

Hedges w/ Floral Planting

Fruit Trees

Raised Planting Beds

Corner Raised Planters at Entry Gate

Benches

FOCH BLVD

165TH AVE

SCALE 1'= 3/32"

ABOVE: A rendering of the garden's interior showing the bright-blue shipping container that functions as a tool shed.

LEFT: One entry to the garden; the elevated rail lines run the full length of the site's longest edge.

OPPOSITE: Section of the rainwater-gathering cisterns.

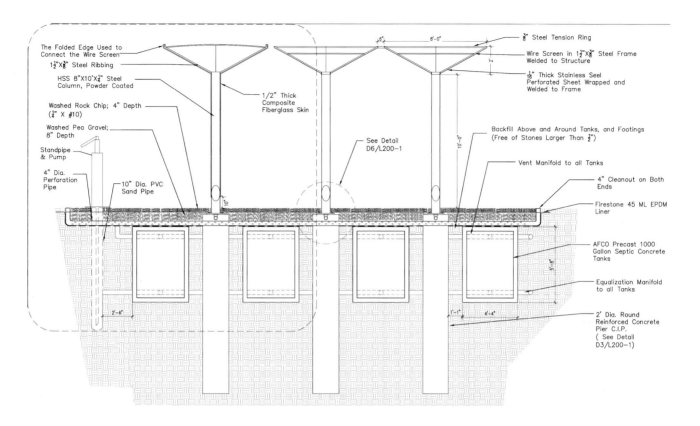

The distinctive blue of the cisterns is repeated again on the painted tool shed, a used shipping container. The container is enclosed by a wooden fence, and the same wood is repeated as the enclosure for discreetly located compost bins as well as in the raised garden beds situated over a bright-green swath of grass. These long, rectangular beds echo the geometry of the nearby rail lines so clearly visible from the site, adding meaning to their context. Wooden picnic tables and benches encourage neighbors to linger. A long row of high wooden frames featuring inset, lightweight metal hoops runs along the edge of the garden, outside its boundary fence, to create a visual invitation to the garden that can be seen from down the street. The frames act as an arbor for climbing vegetables, provide shade, and enhance the sense of place.

Fall plantings selected for the garden's opening season underscored the design intention of attracting the neighbors with a playful design while at the same time providing a functional productive space. Large cabbages were placed decoratively as showy borders for the large, curvilinear beds of brilliantly colored flowers. Wildflowers and annuals were chosen in colors that complemented the intense blue of the rain catchers and shed as well as the bright-red water pump. As an ensemble, the design erases the distinction between the edible and the ornamental. It demonstrates that a plant can be appreciated for its form as well as its nutritional content and, conversely, that a functional shed, pump, bench, or water collector can be considered a beautiful detail instead of an obstacle to beauty.

This project is only one of the many community gardens rescued by the New York Restoration Project (NYRP), an organization founded by Bette Midler in partnership with the Trust for Public Land. Together they rescued 114 community gardens from the fate of being sold to developers by the New York City government in 1999, particularly in under-resourced neighborhoods throughout the five boroughs.[13] The New York Garden Trust was established as a subsidiary of NYRP to take care of fifty-five of the rescued and revitalized gardens, and works closely with the communities and their gardeners.[14] While NYRP's work continues to benefit neighborhoods all over the city, this garden is a stellar example of how design can attract attention and, ultimately, community participation.[15] This project shows that a community garden can heighten food security while enhancing a neighborhood's quality of life in many ways.

ARTSCAPE WYCHWOOD BARNS

DU TOIT ALLSOPP HILLIER | DU TOIT ARCHITECTS, ARTSCAPE, THE STOP COMMUNITY FOOD CENTRE
TORONTO, ONTARIO, CANADA

Adaptive reuse projects strive to maintain connections to the past while conserving resources through the reuse of materials. Depending on location and use, they can also benefit communities by revitalizing neighborhoods. The Artscape Wychwood Barns project in Toronto accomplishes all of these goals. The Wychwood Barns were built early in the twentieth century—the oldest dates to 1913—as repair and maintenance facilities for Toronto's streetcars, and functioned in that capacity until the 1980s. Now a multiuse cultural space with a substantial urban agricultural component, the design incorporates event spaces, fifteen artists' studios, twenty-six live-work studios, and office space for eleven nonprofit arts and environmental organizations in 60,000 square feet.[16]

The design strategy and renovation for this adaptive reuse project focused specifically on allowing many of the complex's original features to be clearly viewed and easily understood. Historic photos, exposed structural elements, and informational plaques tell the history of the barns. Purpose-built adaptations for the various functions outlined above reveal the possibilities of dovetailing the goals of sustainable design with heritage initiatives and new programming needs.

Because of the sensitive, sustainable approach taken by the architects, engineers, and clients, the complex became the first heritage project in Canada awarded LEED certification.

Artscape Wychwood Barns evolved through a partnership between Artscape, a nonprofit development corporation, and The Stop Community Food Centre, a nonprofit organization dedicated to enabling food security in Toronto. Since 2001 these groups have worked together to rescue the barns from the wrecking ball; some neighbors wanted the huge structures to be removed to form an expansive, open park, while others, supported by the City of Toronto and City Councillor Joe Mihevc, as well as heritage associations, helped The Stop and Artscape to keep the initiative alive.[17] The president and CEO of Artscape, Tim Jones, remarked: "We knew at the outset that we would need a strong and compelling vision to attract the level of interest and investment required to bring the Barns back to life."[18] This vision was supplied by architect Joe Lobko and his firm Joe Lobko Architects, now merged with the Toronto firm of du Toit Allsopp Hillier | du Toit Architects, as well as their clients, Artscape and The Stop. Crucial funding from the Metcalf Charitable Foundation and the Ontario government helped to push the project forward.[19]

The Stop is an organization that has been actively providing healthy food for

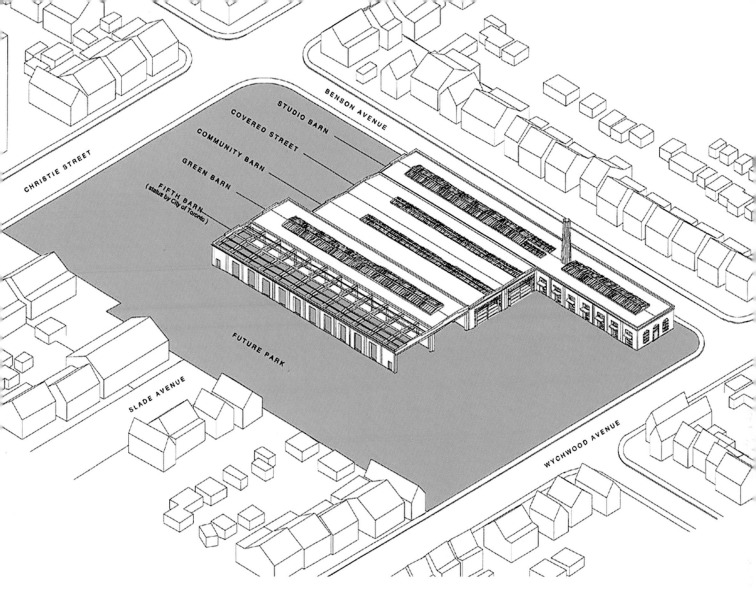

CHRISTIE STREET

BENSON AVENUE

STUDIO BARN
COVERED STREET
COMMUNITY BARN
GREEN BARN
FIFTH BARN
(status by City of Toronto)

FUTURE PARK

SLADE AVENUE

WYCHWOOD AVENUE

ABOVE: The refurbished barn site set in the surrounding park and neighborhood. **OPPOSITE:** Market day in the covered street. Historic photos showing streetcars in Toronto relate the current space to its previous function.

a variety of clients for almost thirty years in Toronto's Davenport West neighborhood. Its mission is "to increase access to food in a manner that maintains dignity, builds health and community, and challenges inequality."[20] Wychwood Barns presented an opportunity for The Stop to adapt its Community Food Center (CFC) model for a new area, the St. Clair and Christie neighborhood. A major tenant there, it occupies one full barn, the Green Barn, a sustainable food production and education center. Its most notable feature is a 10,000-square-foot greenhouse designed for sustainable year-round food production. The greenhouse is a state-of-the-art facility with computer-controlled windows for venting, drip watering systems, and a design that maximizes natural lighting. A compost area, an industrial kitchen that opens onto a gathering/event space, a sheltered outdoor court containing a masonry bake oven, fruit trees, sensitive larger plants, and The Stop's offices fill the remaining space.

In the Green Barn, cooking classes; workshops on growing, composting, and preparing vegetables; and after-school programs foster community

engagement and empower citizens to eat a more healthy diet. The after-school program held within these lively, bright spaces engages 10- to 12-year-old children in growing and cooking activities. They learn how to start a garden and are encouraged to adopt healthy eating habits, but they also study, play games and create art. Evening and weekend workshops include fundraising cooking classes with master chefs, movie nights with food and gardening themes, book launches, bread-baking workshops, and fundraising concerts. The wood-burning oven in the open courtyard is particularly popular for make-your-own pizza gatherings where community members are provided with dough, cheese, and toppings from the community garden and are encouraged to socialize and learn about incorporating vegetables and whole grains into favorite foods.[21]

Sustainable initiatives are woven into the design of the entire Artscape Wychwood complex. Water for greenhouse irrigation and washrooms comes from a large cistern that collects rainwater from the roofs. An effective energy strategy combines a ground-source heat pump HVAC system with

![plan]

Benson Avenue

Barn 1

Barn 2

Barn 3

Barn 4

Barn 5

18

Christie Street

Wychwood Avenue

22

19

20

TOP: The plan of the barns and surrounding park.
ABOVE: The Stop greenhouse soon after completion, showing some of the new plantings.
OPPOSITE, ABOVE: The design concept for The Stop sheltered garden and community bake oven.
OPPOSITE, BELOW: The design concept for The Stop greenhouse.

passive approaches to energy conservation, lighting strategies, air quality, and thermal comfort. In addition, Artscape negotiated with the city for a variance that allowed fewer parking spaces than the zoning called for to provide more orchard space and to encourage users to walk, bike, or take public transport.

The oldest of the barns now functions as a covered arcade linking artists' studios with offices and The Stop's greenhouse facility. On Saturday mornings in the cold months, this "street" hosts The Stop's large farmers' market. Hundreds of shoppers come to buy local produce, locally roasted organic coffee, heritage organic seeds for growing their own herbs and vegetables, and artisanal cheeses, breads, and preserves. In the warm months, the market takes place in the surrounding park. An outdoor playground makes shopping at the farmers' market a family activity and helps children associate healthy food with fun.

Artscape Wychwood Barns helps the community by providing access to nutritious staples and programs that enable food production and processing, and has become a hub for social exhange. Community engagement was central to the redesign of the Wychwood Barns and led directly to its success. This thriving example of the CFC model led The Stop to conduct a study in 2010 that examines the possibility of creating a web of CFCs in cities across Ontario.[22]

EVERGREEN BRICK WORKS

DU TOIT ALLSOPP HILLIER, DIAMOND+SCHMITT, ERA ARCHITECTS,
CLAUDE CORMIER ARCHITECTES PAYSAGISTES, EVERGREEN, AND
THE TORONTO AND REGION CONSERVATION AUTHORITY
TORONTO, ONTARIO, CANADA

Former industrial sites are increasingly being transformed into community spaces, cultural institutions, and social service centers, and adaptive reuse strategies are evolving to preserve heritage and take advantage of the existing structures' resources. Urban agriculture is also being incorporated into these developments more frequently, as Evergreen Brick Works in Toronto, an exemplary project, reveals. Located in the city's former Don Valley Brick Works, this ecologically sensitive, 40-acre site spread across a ravine is deeply embedded with Toronto history. In 1889 a clay quarry and factories were established here to manufacture brick for the growing city's new construction projects, including the historically significant Old City Hall.

In 1994 this large site became part of Toronto's park system after the factories and quarry were closed. The adaptive reuse of the Brick Works and its extensive landscape is an initiative of Evergreen, a nonprofit organization working in partnership with the City of Toronto. The project lead was Joe Lobko of du Toit Allsopp Hillier | du Toit Architects Ltd., it was designed by architectsAlliance, and includes landscape architecture by Claude Cormier Architectes Paysagistes.[23] Evergreen created a Neighborhood Engagement Committee to help connect Toronto's communities with the resources available at the Brick Works, provide advice, and give feedback on the opportunities for and barriers to community participation.

As part of its mission of addressing both environmental and community health issues, programming incorporates edible landscaping and a farmers' market to bring to the community local, fresh produce and artisanal foods such as cheeses, breads, and jams. Central to the initiative are demonstration areas to teach about urban agriculture that include 110,000 square feet for a vegetable garden and nursery.

Due to their dilapidated condition, some of the original buildings could not be reused, but their components were carefully stacked and stored for reuse in

ABOVE: A rendering of the renovated Brick Works site.
RIGHT: A historic photo of the original Don Valley Brick Works.
FAR RIGHT: A photo of the Evergreen Brick Works site before the renovation.
LEFT, ABOVE: A rendering of the Centre for Green Cities by Diamond+Schmitt Architects.

other construction projects. In addition, four of the existing buildings on the site required a new floor that could withstand substantial loading; an innovative flooring was designed from a recycled plastic material called Cupolex, a strong and lightweight substrate that was laid beneath reinforcing bars and covered with only a thin layer of concrete. This strategy significantly reduced floor weight while enabling the heavy loads required by garden facilities.

Visitors to the complex encounter the Welcome Court first, which features a sheltered garden of thirteen different beds planted in an area formerly paved over with concrete—now visitors feast their eyes on a variety of edible plants, native trees, shrubs, wildflowers, and marsh plants. The adjacent Welcome Center provides general and special event information and serves as a meeting point for tours and educational activities. Tours include a guided walk through the 52,000 Kilns Building's restored brick kilns and drying tunnels.

The Pavilions, a 27,000-square-foot sheltered outdoor space made from one of the former factory buildings, houses farmers' markets, plant nurseries, and festivals. The Chimney Court is a play area for children that was designed with edible landscaping, including fruit trees and berry bushes. The Ruin Court,

made from remnants of a building that was beyond repair, features industrial artifacts, including antique machinery unearthed in the construction process.

A nursery building, Evergreen Gardens, includes demonstration gardens for edible plants and native species, and is also a resource where people can buy plants; seeds; organic soils and fertilizers; bird, bat, and butterfly houses; and learn about growing vegetables organically, choosing beneficial nonnative plants, and raising a variety of heirloom vegetables.

A second market, the Evergreen Marché, provides access to other local farmers and producers and is also a place to learn about cooking and gardening. It is adjacent to the Discovery Gardens, a factory building that was transformed into a sheltered, open-air, native-plant demonstration garden that teaches school and community groups about plant care and maintenance. Its new Cupolex floor supports 20,000 square feet of native plant demonstration beds in the form of large, raised mounds. In the winter, this active space is completely transformed as the planted mounds become ornamental objects and the space becomes an ice-skating rink.

The Centre for Green Cities, designed by Diamond and Schmitt Architects, is the only new structure at the Brick Works, conceived as a multifunction building with classrooms on the second level and offices for the Evergreen Foundation on the upper floors. It is conceived of as an "ideas incubator."[24] Natural ventilation with thermal chimneys keep the building cool in summer. A solar cogeneration system provides electricity. Storm water is collected in enormous 5,000-gallon cisterns and then used in cooling towers on the roof, as well as for graywater

purposes and irrigation. The building's most distinctive features are the movable screens for solar shading on the building's facade and window boxes where tenants can grow plants into a vertical wetland that will naturally filter rainwater. Large-scale artwork also clips onto the building skin. In addition, composting and recycling systems reduce tenant waste. The goal is to achieve a performance of 58 percent less energy consumption than a conventional office building. The Centre is also targeting a LEED Platinum designation for this design.[25]

A scenic overlook, the Ravine Terrace, provides views of the site's historic quarry, wetlands, and ponds. A separate Arts and Adventure Centre provides space for artists' workshops and art classes, and offers bicycle, snowshoe, and cross-country ski rental so people can experience the vast site in every season. A nature-based children's playground will teach an appreciation of nature while promoting good exercise habits.

This project integrated urban agriculture from its inception as one of the core means for obtaining community engagement as well as a way to promote healthy eating and lifestyles. Together with the farmers' market and other health-conscious programming, the Evergreen Brick Works has succeeded in increasing food security and bringing people together.

ABOVE: A rendering of children playing in the Chimney Court. Edible landscaping—fruit trees and berry bushes— appears in the center. OPPOSITE, ABOVE: The Welcome Court, featuring a garden that includes edible plants, native trees, shrubs, wildflowers, and marsh plants. OPPOSITE, BELOW: The master site plan. Areas dedicated to urban agriculture or food are represented with superimposed vegetables and artisanal products.

ABOVE: Native plant beds in the forms of blooming mounds during the summer in the Discovery Gardens building.

RIGHT: The Discovery Gardens converted to an indoor ice rink for the winter months.

RIGHT, BELOW: A rendering of the Welcome Center; many of the site's original industrial artifacts are incorporated into the design.

BELOW: A detail of the living wall showing its scale and the materials used on the facade.

OPPOSITE: A detail of the living wall on the exterior of the Centre for Green Cities; movable screens are designed for solar shading, and window boxes full of plants create a vertical wetland to filter water.

NIAGARA COMMUNITY FOOD CENTRE
JORDAN KEMP EDMONDS
TORONTO, ONTARIO, CANADA

The Niagara Community Food Centre is an exploration of the Community Food Center (CFC) idea from an architectural perspective. CFCs provide opportunities for the use of undervalued and wasted space to foster knowledge and engagement about food, and of course produce healthy food as well. This project imagines the type of space and structure that would be necessary for food and agriculture within an urban community. It proposes that CFCs can act as catalysts for a revitalization of our urban ecosystems, reenvisioning urban parklands as places of intersection between community and agriculture and as a celebration of the local ecological and industrial heritage. This particular project is intended as a prototype for a network of future urban landscapes that would focus on creating a sense of community around the need for local food production.

This proposal is for a specific site in the Niagara neighborhood of Toronto, on a parcel that runs adjacent to one of the city's historic rail corridors and that is also intersected by Garrison Creek, an ancient tributary of Lake Ontario that was buried beneath the city many years ago. The site is centered among a number of brownfield lands that served historically as agro-industrial sites for icehouses, slaughterhouses, lumberyards, and farm equipment manufacturers, some of which still exist. It is a neighborhood with a rich and colorful industrial and ecological past that is undergoing an intensive and fast-paced transformation into a new high-density residential community due to its proximity to downtown Toronto. The Niagara CFC engages this heritage, protecting and enhancing a connection with the past while framing a new perspective on food, agriculture, and urban community for the future.

The site's value as a new landmark within the city is strengthened by the creation of a green pathway axis, which connects the new Garrison Commons parklands with the new West Toronto Rail Path. This interweaving of community green space is fundamental to a CFC, which seeks to normalize food and agriculture as the commonly understood language of what is included in an urban park. This direct connection to large, open green spaces enables a CFC to encourage the long-term development of productive community

gardens and edible landscapes. Here the CFC plays a key role as steward to the park, serving to manage and facilitate community garden spaces, and providing key infrastructure and education to its community gardeners.

The project adopts principles of sustainable construction, including secured access to solar and water resources, reduced embodied energy, and adaptive and healthy architectural environments. The program includes four primary spaces: a Community Hub, an Urban Barn, the Niagara Community Greenhouse, and the Garrison Commons Community Gardens.

The Community Hub is the programming core of the structure and an interface with the Niagara neighborhood, providing a street presence that promotes accessibility and inclusivity for local residents. The Hub houses a community kitchen and café, a resource and information booth, and office and meeting spaces for food-related, not-for-profit services and grass roots community business enterprises.

The Urban Barn houses the main event space, a flexible environment that is comfortable during all seasons. Here the community can gather to discuss issues related to food and agriculture and spread cultural knowledge about food systems. A barn door folds open to the surrounding terrace, turning the building into a floating canopy for farmers markets, food fairs, and other outdoor community events.

A large Community Greenhouse enables gardening, education, and agricultural research throughout the year, maximizing harvests to ensure a sustained local supply of fresh produce for the Hub's community meals and food bank programming. The greenhouse environment is conditioned by solar thermal energy and a geothermal storage system, activated by a large thermal store that doubles as the space's circulation spine. A large cistern fed by rainwater from the greenhouse rooftop irrigates the gardens.

The Garrison Commons Community Gardens flow out from the greenhouse and into the park, expanding available growing space for the greater community. Public garden plots would be assigned by the CFC staff, who would foster community gardening while enhancing and protecting the parklands.

The Niagara CFC anticipates a future when local food production may be a necessity due to the scarcity and cost of fossil fuels and the need to reduce greenhouse gas emissions. However, its purpose extends beyond environmental responsibility—it seeks to convert the demand for local food production into a vehicle for strengthening local community ties. It proposes a vision for a future when CFCs become a cornerstone of urban neighborhoods. These will help nourish cities in the physical sense as well as the cultural sense, by bringing people together via their food, their ecosystem, and their local heritage.

THE SCIENCE BARGE
NY SUN WORKS AND GROUNDWORK HUDSON VALLEY
YONKERS, NEW YORK

Experimentation and education are often combined goals of urban agriculture projects. Teaching current and future generations about the potential for food production in urban areas is thus often intertwined with the development of cutting-edge techniques. The Science Barge is a perfect example of this double purpose.

The Science Barge is a prototype sustainable urban farm on a 400-square-meter barge in Yonkers, New York. It was conceived in 2007 by the environmental nonprofit organization NY Sun Works as an education center for students learning about local food production, alternative energy sources, and water management, allowing the organization to experiment with all three.

The focus of the project is to use renewable energy for food production with zero net carbon emissions, zero chemical pesticides, and zero runoff. Solar panels, wind turbines, and biofuels generate all the energy needed to power the barge, and the greenhouse is irrigated solely with collected rainwater and purified river water; it operates completely "off the grid." It is the only fully functioning demonstration of renewable energy supporting sustainable food production in the New York area.

The greenhouse features several examples of recirculating hydroponic systems that reduce land and water use. It also helps reduce the carbon, particulate, nitrogen, and sulfur oxide emissions that result from truck deliveries by lowering the farm-to-table distance. To avoid pollutants inherent in chemical pesticides, the Science Barge uses insect-based integrated pest management. The project currently grows tomatoes, cucumbers, melons, strawberries, lettuces and leafy greens, beans, and herbs. NY Sun Works claims that if all the fallow rooftop spaces of New York were fitted with techniques similar to those employed on the Science Barge, the city could produce enough vegetables to feed its entire population.

The Science Barge relies entirely on renewable energy. The base source is a set of solar photovoltaic panels, but it also houses wind turbines that were carefully selected for urban settings—they cope with fluctuating wind conditions and minimize noise disturbance. Biodiesel produced by waste vegetable oil from agricultural and food industry byproducts is also used to generate power. Thirty-eight million liters of waste oil are generated in New York

ABOVE: A view of the Science Barge interior, showing hydroponic techniques in development.
LEFT: Composting in progress.
OPPOSITE: The barge docked in the Hudson River on the Yonkers waterfront, its permanent home.

ABOVE: Hydroponic plants and water-collection tanks used to gather rainwater.
ABOVE, RIGHT: A view of the hydroponic system's structure, showing recirculating water.
OPPOSITE, ABOVE: WindowFarms, a hydroponic watering system made from recycled plastic bottles.
OPPOSITE, BELOW: The Science Barge, shown here in its original mooring in New York City, holds active greenhouses, photovoltaic panels, and outdoor seating.

City restaurants that could be converted to biodiesel.

Agriculture consumes two-thirds of global fresh water, and its runoff often contaminates groundwater sources. The Science Barge addresses this by adopting closed-loop systems for water use. Once purified river water or collected rainwater enters the system, the drainage water from the recirculating hydroponic systems is collected for reuse, and any runoff water from the farm is used for the constructed wetlands system on board. As a result, there is no discharged water.

Originally moored in the East River, the Science Barge has found an entirely new life a few miles upstream from Manhattan. It was donated in October 2008 for the symbolic sum of two dollars to Groundwork Hudson Valley, a leading environmental organization in Yonkers, a suburb north of New York City, and has become the centerpiece for the redevelopment of its downtown waterfront. Groundwork Hudson Valley has maintained most of the basic components of the original Science Barge, and has reinforced and streamlined its educational function by developing curricula adapted to various school levels as well as the general public. Moreover, the experimentation with growing techniques has continued.

The Science Barge has already inspired a surprising number of similar projects and enabled NY Sun Works to

establish itself as a leader in adapting methods of intensive food production to urban settings. BrightFarm Systems, a commercial enterprise that specializes in designing systems for commercial and educational hydroponics, was created as the for-profit arm of NY Sun Works, and builds directly on the lessons learned on the Science Barge. Similarly, Jenn Nelkin and Viraj Puri, two members of the Science Barge team, cofounded Gotham Greens, a commercial rooftop hydroponic farm that raised over one million dollars in capital from private investors and state agencies. A second barge from New York, the Waterpod, also functioned briefly as a one-season experiment in 2009 in communal living for artists and included urban agriculture components, such as chicken and vegetable beds, along with rainwater catchment, solar heat, and composting.

The history of the many ways the Science Barge has continued to develop beyond the purposes of its original mission shows how a project that seeks to educate about urban agriculture can serve to fulfill multiple other functions: connecting food production and resource use, serving commercial expansion, even urban redevelopment. It is thus a good example of the many functions urban agriculture can serve and the many benefits that can be provided to communities that incorporate it.

THE EDIBLE SCHOOLYARD
WORK ARCHITECTURE COMPANY, PUBLIC SCHOOL 216, AND EDIBLE SCHOOLYARD
BROOKLYN, NEW YORK

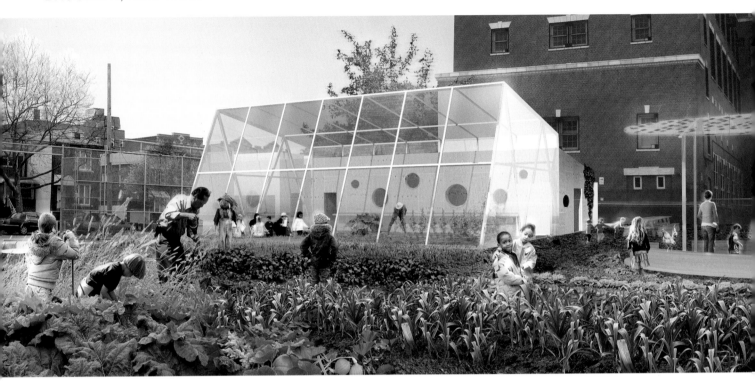

Schools offer a unique opportunity to initiate changes of attitude regarding the provision of food within an urban context. Several celebrity chefs such as Jamie Oliver have begun campaigns to improve the quality of school meals and change how schools teach about food. The Edible Schoolyard program, launched by renowned chef and author Alice Walters and the Chez Panisse Foundation in 1995, aims not only to improve school meals, but to give schoolchildren and their communities the actual skills they need to grow and prepare local, healthy, and nourishing food. It also teaches about the environmental benefits of urban agriculture. Good design has been integral in the development of the Edible Schoolyard from the start.

The Edible Schoolyard project began at the Martin Luther King, Jr. Middle School in Berkeley, California, where a 1-acre organic garden and kitchen was created on an adjacent vacant lot. This acts as an interactive classroom used by teachers and specialist educators to integrate food systems into the core curriculum. The project has been successful in raising awareness about food issues in the Berkeley community—many other schools now have productive gardens. It was also instrumental in the overhaul of the local school lunch program. Since 2005 the Chez Panisse Foundation has been developing partnerships with like-minded organizations throughout the United States to demonstrate that the Edible Schoolyard principles are applicable elsewhere. The first Edible Schoolyard project in New York is being established in what was the parking lot at P.S. 216, the Arturo Toscanini School in Brooklyn, which educates about 470 children from kindergarten to fifth grade.

Edible Schoolyard NY is being designed by WORK Architecture Company, which often integrates urban agriculture elements into its projects. The challenge for the architects was to create an environment that could accommodate a comprehensive, interdisciplinary curriculum tied to state learning standards and that would connect food to academic subjects taught at school. The resulting unique learning environment features three specially designed spaces that complement a quarter-acre productive organic garden to create learning, growing, and cooking areas that can function throughout all four seasons. A Kitchen Classroom is surrounded by a mobile greenhouse on one side, and the multifunction Systems Wall on the other contains various service functions that ensure the building's self-sufficiency. In the garden, a structure known as the Ramada serves as a gathering point for external activities.

TOP, RIGHT: Communal tables and the school's outdoor oven.
RIGHT: An outdoor class being held in the Ramada.
OPPOSITE: The greenhouse extended its full length to protect crops in the early fall.

The Kitchen Classroom, at the core of the project, offers space for food preparation and consumption, and shares characteristics with large farm kitchens that also act as social spaces. A continuous counter wraps the perimeter, and three learning stations equipped with appropriate kitchen equipment and storage facilitate child-friendly food preparation. Portholes provide views into the adjacent greenhouse. In the center, three large dining tables allow up to thirty students to eat together in a communal format that encourages social interaction, debate, and an appreciation of food.

The Systems Wall, built directly into the structure's envelope, is designed to assist in many of the building's service functions. This series of round spaces accommodates a variety of technical functions within the building: a 1,500-gallon reclaimed water cistern stores rainwater harvested from the sloping roof for uses within the building or garden, including in the adjacent dishwashing station; a waste station provides a location for collection of waste and separation into useful waste streams for reuse, composting, or recycling; a chicken coop, storage for farming tools and other equipment, and batteries that store harvested solar power are also included. Working together as a series of interlinked sustainable systems, these elements create an infrastructure intended to be entirely off-grid.

The Mobile Greenhouse opposite extends the school's growing season. It covers 1,600 square feet of growing soil during the fall, winter, and early spring to provide sheltered growing space. In the summer months, this lightweight polycarbonate structure simply slides over the kitchen, exposing the earth to the open air.

The Ramada in the organic garden extends many of the kitchen activities to the outside. An outdoor oven bakes meals in the summer while the children tend the surrounding gardens. A large picnic table and a round seating area provide a location for children to gather and receive instructions for lessons before starting the day's farming and cooking activities. The whole structure is shaded and protected by a roof of photovoltaic panels that feed electricity into the battery store in the Systems Wall.

This project engages schoolchildren, their parents, and the community in the process of food production, but also demonstrates the principles of self-sufficiency. Regular discussions focus on how the food system impacts health, nutrition, and the environment. High-quality design is instrumental in bringing these concepts to a wide audience and reveals the potential for urban agriculture to enrich the educational experience and the learning environment.

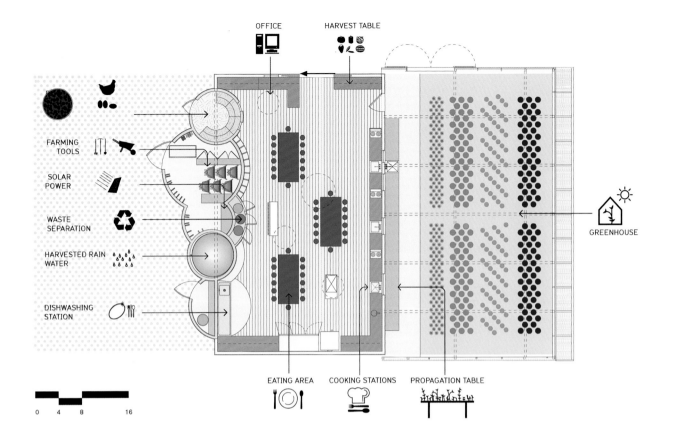

OFFICE

HARVEST TABLE

FARMING TOOLS

SOLAR POWER

WASTE SEPARATION

HARVESTED RAIN WATER

DISHWASHING STATION

GREENHOUSE

0   4   8   16

EATING AREA

COOKING STATIONS

PROPAGATION TABLE

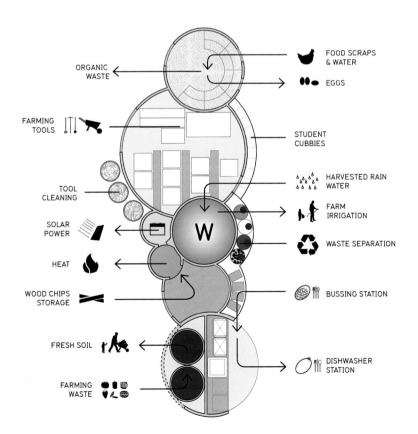

ORGANIC WASTE

FARMING TOOLS

TOOL CLEANING

SOLAR POWER

HEAT

WOOD CHIPS STORAGE

FRESH SOIL

FARMING WASTE

FOOD SCRAPS & WATER

EGGS

STUDENT CUBBIES

HARVESTED RAIN WATER

FARM IRRIGATION

WASTE SEPARATION

BUSSING STATION

DISHWASHER STATION

ABOVE: A plan of the Systems Wall, Kitchen Classroom, and greenhouse.
LEFT: Functions performed by the Systems Wall in detail.
OPPOSITE, TOP: Crops growing in the fully extended greenhouse in midwinter.
OPPOSITE, CENTER: Children tending crops in the organic garden; the greenhouse is retracted over the Kitchen Classroom building for the season.
OPPOSITE, BELOW: The multifunction, internal learning environment.

THE EDIBLE CAMPUS

McGILL UNIVERSITY, ALTERNATIVES, AND SANTROPOL ROULANT
MONTREAL, QUEBEC, CANADA

The Edible Campus vegetable garden appears each spring at McGill University in Montreal. It began in 2007 as a collaborative action/research project between the Minimum Cost Housing Group at McGill University and two nonprofit organizations—Santropol Roulant and Alternatives.[26] The team of academics and NGOs was interested in growing food in the heart of the city through the cultivation of edible landscapes in order to increase food security. In 2007 one person in six was food insecure in Montreal.[27] This project developed innovative methods for gardening in challenging city spaces, increased access to healthy food, and reduced food miles; moreover, it developed strategies for making cities more sustainable by decreasing the heat-island effect caused by hardscape surfaces.

A shortage of community garden plots in Montreal highlighted the need for more growing space in the city, particularly in its center. This problem was magnified by the discovery in 2007 that the soil of many community garden sites in the city—often located on former brownfield sites—was contaminated beyond safe levels for food production. Although urban land is extremely expensive, the group found many "overlooked, under-utilised, and neglected areas" just waiting to be used creatively.[28] Hardscapes, such as concrete and masonry plazas as well as walls and roofs, are often the only available open surface, so the group set out to use these resources. Imaginative approaches to container gardening turned out to be an economical way to turn a concrete surface into a growing surface. Purpose-built planters with built-in reservoirs for ease of watering, called sub-irrigation planters, were designed for these locations. As small mobile units, they can be configured in infinite ways and removed in the winter, and plants can be started in the planters indoors before the short Montreal growing season begins.

After garden design considerations were addressed, including identification of potential growing surfaces, daylight/shadow studies, identification of water resources and tool storage areas, and development of the planter with its reservoir, the group chose a very visible, public site as the prototype garden to reduce the possibility of vandalism, make it as noticeable as possible, and increase the enjoyment of the garden by all. Several dozen planters were placed on flat concrete areas and some unused

steps. Arbor hoops and ropes for plants to climb on made the vegetables showy, decorative additions to the urban landscape, as well as food. What was once a bleak concrete expanse became an explosion of greenery. In addition, the garden demonstrated the possibilities of landscape interventions to students, neighbors, and passersby. "It engages, involves, and includes the citizenry to consider aspects as mundane as the design, composition and orientation of the containers."[29]

Originally twenty-five plant species were selected for their forms, scents, and biodiversity. The crops included arugula, basil, bok choy, broccoli, cantaloupe, celery, cherry tomatoes, chives, cilantro, cucumbers, dill, edible flowers, eggplant, green beans, green peppers, ground cherries, lettuce, leeks, mint, onions, parsley, squash, Swiss chard, Thai basil, and toma-toes.[30] The following year, 2009, the Edible Campus program expanded with the addition of a 100-square-meter raised-bed rooftop garden.

The Minimum Cost Housing Group depended on the other partners to maintain the garden through the hot summer months. Because Alternatives and Santropol Roulant are staffed by volunteers, they had the human resources to tend the numerous tiny plots clustered on the campus. Volunteers planted, weeded, watered, and harvested. During the height of the 2008 growing season, one-third of Santropol Roulant's produce needs for their meals-on-wheels program, over 375 pounds, came from this single urban garden.[31] Produce was carried by foot or bicycle to the Santropol kitchen, where healthy meals were prepared and delivered to mobility-impaired Montreal residents. Food waste was also captured from Santropol Roulant's meal-preparation activities, composted, and then used back in the garden.

The variety of citizens involved in the project was the key to its success. The garden was recognized with a 2008 National

TOP, LEFT: Container gardens with added arbors allow plants to grow into arches.
TOP, RIGHT: Plants climbing the concrete facade of an academic building on campus.
ABOVE: Plants add lushness to a formerly all-concrete plaza.
OPPOSITE: Raised vegetable beds and container plantings transform a formerly under-utilized rooftop into a productive landscape.

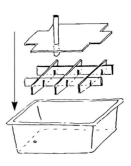

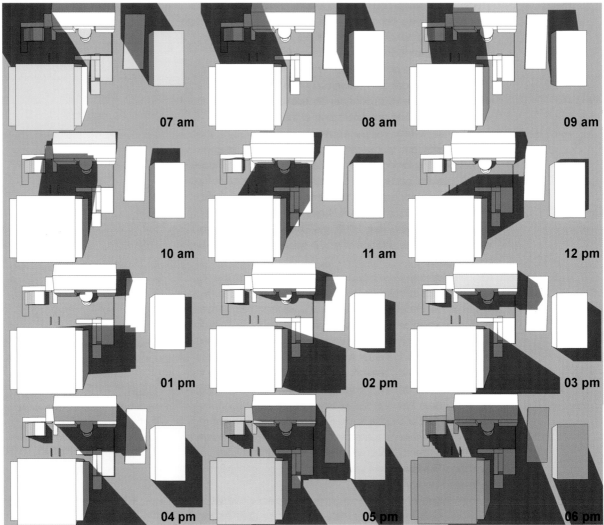

07 am

08 am

09 am

10 am

11 am

12 pm

01 pm

02 pm

03 pm

04 pm

05 pm

06 pm

ABOVE: Volunteers tending the campus's container gardens.
OPPOSITE, TOP LEFT: An aerial view of the main garden space in the process of being developed.
OPPOSITE, TOP RIGHT: A design sketch for the innovative sub-irrigation planters featuring built-in reservoirs that enable the mobile containers to be used on any hardscape surface.
OPPOSITE, BELOW: A sunlight study to determine when plants will receive the strongest sunlight; this informed the positioning of planters and various species.

Urban Design Award, and the design jury indicated an appreciation for how the Edible Campus was "creating a sustainable prototype that could potentially be expanded to other university campuses and across the city."[32] Other campuses are indeed duplicating this model. The University of Quebec in Montreal now has aesthetically pleasing urban agriculture plots on its campus, Hart House at the University of Toronto began an urban agriculture initiative in 2009, the Massachusetts Institute of Technology now has what they have dubbed an "EarthBox Garden" composed of sub-irrigated planters, and in 2010 fruit trees were planted at Allegheny College in Pennsylvania, to name a few. The great potential for food production on neglected and underused spaces—including on concrete, the most inhospitable of surfaces—is the most significant lesson provided by the Edible Campus.

WOOD STREET URBAN FARM
GROWING HOME AND SHED STUDIO
CHICAGO, ILLINOIS

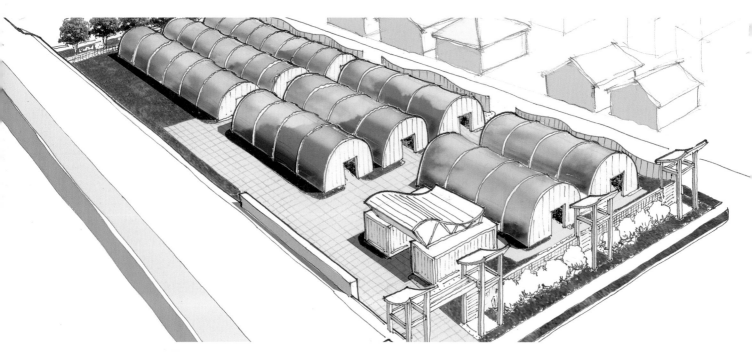

Growing Home operates a transitional employment program for Chicagoans who face barriers to entering the workforce from homelessness, prior incarceration, or substance abuse. It offers participants six months of farming employment while training them in general job-readiness and organic agriculture skills. Growing Home's late founder, Les Brown, said: "Homeless people often are without roots. They're not tied down, not connected, not part of their family anymore. Our organic farming program is a way for them to connect with nature . . . When you get involved in taking responsibility for caring for something, creating an environment that produces growth, then it helps you to build self-esteem and feel more connected." Nearly 90 percent of Growing Home's employees end up finding stable housing, and two-thirds of them go on to either full-time jobs or further job training.[33]

The origins of Growing Home date to 1992, when the Chicago Coalition for the Homeless acquired federal surplus land near Navy Pier to start an urban farm. Four years later, the program was launched.[34] Growing Home currently operates three farm sites, all of them certified organic: a 10-acre site in Marseilles, Illinois, and two urban farms on Chicago's South Side. One of these latter sites,

the Wood Street Urban Farm, has been the subject of a multiyear collaboration between Growing Home and a pair of local architects, Rashmi Ramaswamy and Mike Newman of SHED Studio, who are committed to sustainability, social justice issues, and innovation in design. Ramaswamy has been involved in Growing Home since the mid-1990s and also served on the board of the Chicago

Coalition for the Homeless, where Growing Home had its roots.

The Wood Street Urban Farm is located in Englewood, a once-flourishing Chicago neighborhood. However, since the early 1960s, the area has suffered from decades of neglect and the flight of about half of its population, leading to extensive abandonment and resulting expanses of vacant land. This cycle of decline has allowed drug trafficking and other criminal activity to thrive. As a result, most major food vendors have abandoned much of the South Side of Chicago, resulting in a classic "food desert."

Growing Home staff, led by Harry Rhodes, participated in a 2005 planning process to improve the neighborhood. One of its goals was to "develop an urban agriculture district to provide business, job training and employment opportunities while improving the availability of fresh produce," combining strategies for economic development and healthy living into an initiative

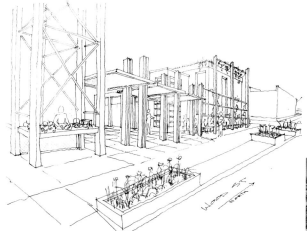

ABOVE, LEFT: Sketch for revised Urban Farm project, with larger building along Wood Street, May 2007.
ABOVE: View of the farm in 2008, with containers still in use as support spaces.
LEFT: The opening ceremony of the realized project.
OPPOSITE: An early site perspective, July 2006.

focusing on access to and production of healthy food. Growing Home's aim of developing community-based urban farms lent itself perfectly to this objective for Englewood. Consequently, Growing Home entered a partnership with several local entities that started work on fleshing out the vision of such an urban agricultural district—with SHED Studio playing an essential role in helping develop this vision.

The original 2-acre site assigned by the local alderman and the City of Chicago in 2005 was ultimately replaced with a smaller site on Wood Street covering two-thirds of an acre. Over the course of the next six months, SHED met regularly with key Growing Home staff to find out their goals for the site, from fencing to farming. The result was a plan for a sustainable urban farm with a multipurpose building.

Land on Wood Street was transferred from the City of Chicago to Growing Home for one dollar, and the city also provided initial funding through a special social enterprise venture grant. The development of the urban farm was thus formally launched in September 2006. The initial phase included finalizing the site plan and building three hoop houses—unheated greenhouses used to grow crops year-round. In total, 8,400 square feet of covered growing space was ready for the summer 2008 season.

To further enhance its mission by adding space for cleaning and packing produce, a classroom, and office space, SHED developed a new building for Phase 2 of the program.

Ornamental landscaping was also added to make the site an amenity for the community. A living wall, flower boxes, and an area for the farm stand were created for the east side of the site, while the western side boasts an "edible fence" with fruit trees and bushes. Green elements that will make the farm energy efficient are planned for the final phase, including solar panels and a "power tower" with photovoltaic panels or a wind turbine and a green roof to insulate the space underneath.

Wood Street now produces over 10,000 pounds of produce a year. As an active participant in the drafting of the city's "Eat Local, Live Healthy" initiative, Growing Home set itself a goal to sell half of what is grown at the Urban Farm within the Englewood community; it sells at the Englewood Farmers' Market as well as its own farm stand, other green markets, and area restaurants.

The longstanding role of SHED Studio in the creation of the Wood Street Urban Farm is a good example of the lengthy collaboration that is often required to develop comprehensive urban agriculture sites. SHED supported negotiations with the city and with contractors. They also sourced many in-kind donations of materials for the building, which substantially lowered the cost. Growing Home gave Newman and Ramaswamy a special award at its 2009 Annual Benefit in recognition of their contribution. This project illustrates well how, for committed design firms, working on urban agriculture can transcend the usual client-architect relationship to become a true partnership.

LEADENHALL STREET CITY FARM
MITCHELL TAYLOR WORKSHOP AND BRITISH LAND
LONDON, UNITED KINGDOM

In many cities around the world, prominent central building sites stand empty for years while developers negotiate with planning departments or wait for financing. These wasted spaces are often boarded up or used as temporary parking lots, offering little to the local community. One such site is 122 Leadenhall Street in the City of London, which is destined to hold a forty-seven-story commercial building by Rogers Stirk Harbour + Partners architects. The project was mothballed in 2008 due to the poor economy and associated reduced demand for office space, so the site owner, British Land, organized an ideas competition for young architects, challenging them to find a temporary use for the site—with a budget of only £125,000.

This 2009 proposal for a city farm by Mitchell Taylor Workshop was selected from a short list of thirteen as the competition winner. The firm wanted to create a contrast to the "sterile and joyless box designed to keep the seasons out" that characterizes many other buildings near the site. Introducing a living environment to the City is a surprising move perhaps, but the architects questioned why the space, although located in London's dense financial district, should not benefit from "the screech of exotic birds, the scent of bougainvillea, the amazing burst of colour in spring from wisteria and flame trees."[35]

The project is full of innovative ideas about how urban agriculture and the commercial world of finance can interact and coexist. Rather than staring at a vacant lot, the occupants of the surrounding buildings who overlook the site would instead be exposed to a changing landscape that reflected the seasons, produced nourishment, and provided an environment for relaxation and contemplation. Lunches made with fresh ingredients grown on-site could be sold at kiosks; locals could also visit chickens or pick up some fresh eggs, vegetables, goats' milk, or berries.

The proposal divides public space into three distinct growing areas designed around the climatic conditions imposed by the surrounding buildings. Vegetables, soft fruits, herbs, and root vegetables that need the most sunlight would be located in the sunny, northern part of the site. Leafy green crops, such as cabbages, broccoli, and spinach that can cope with partially shaded conditions would occupy the central part of the site. The southern area, cast into perpetual deep shade by adjacent buildings, would hold a log forest of exotic mushrooms interspersed with shade-tolerant crops, such as rhubarb and mint. City workers could escape their offices for a shady lunch,

TOP, RIGHT: Wildflowers growing on a slope of infill provide a visual backdrop for the farm's vegetable beds.

CENTER, RIGHT: Areas of the site cast into deep shade by surrounding buildings are devoted to shade-loving produce, including mushrooms and rhubarb.

RIGHT: Flower meadows provide office workers respite from the dense urban surroundings.

OPPOSITE: Economic downturns have caused an indefinite delay in proposed new construction on this site in the City of London.

and visiting school groups could learn about food sources and preparation. The architects even discovered that medieval city planning laws still in effect would allow the farm to keep sheep, goats, pigs, and a cow. A south-facing ramp built over the site's existing mound of infill would be used to grow a spring flower meadow for early-season impact.

Leadenhall Street frontage would be enlivened by a series of vegetable-shaped cutouts peeking through an expressively designed perimeter fence made of recycled construction boards that serve as outlets for selling fresh produce grown on-site and take-away meals prepared with local ingredients to pedestrians as they pass by on the way to and from work. The frontage would also create a public face for the farm.

Permanent infrastructure requirements needed to implement the project would have been minimal. Some terracing would be needed, requiring digging equipment and scaffold boards for retaining devices. The proposed installations are all temporary and low-cost; reuse of materials and components from the existing site is highlighted. Four "vertical growing enclosures" constructed by architecture students are suggested for housing chickens and other livestock, and include a wire mesh cladding to serve as supports for berries and other climbing plants. Planters made of standard-size precast concrete drainage channels and construction bags filled with

soil would be configured in many ways, and would be movable and recyclable.

Management of the project would be overseen by a community-supported agriculture organization. Architect Piers Taylor estimates that the project was achievable within the allowed budget and would have been easy to install or to remove should the market change quickly. The architects' hope for the project was to make office workers more aware of the possibilities offered by urban agriculture, in turn become more supportive of this or similar inclusions in the urban fabric and even demand a new urban environment more closely integrated with the natural cycles and systems around them.

British Land ultimately decided not to pursue this project due to perception concerns about sending the wrong message to the City at a time of economic recovery. While this was a missed opportunity, the proposal has, however, stimulated discussion about how other vacant sites can start to engage the public realm. The City of London's head of planning has encouraged developers to consult with architects about temporary makeovers of recession-hit sites to ensure that they do not cause urban blight. At the site of the demolished Middlesex Hospital in Fitzrovia, a campaign by locals to provide temporary allotments and educational activities that would make the space available to the community was successful.

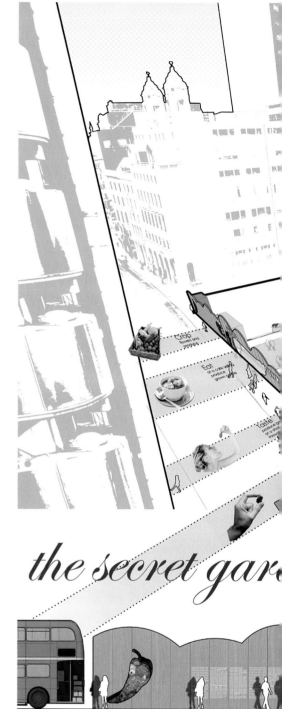

LEADENHALL STREET
ELEVATION

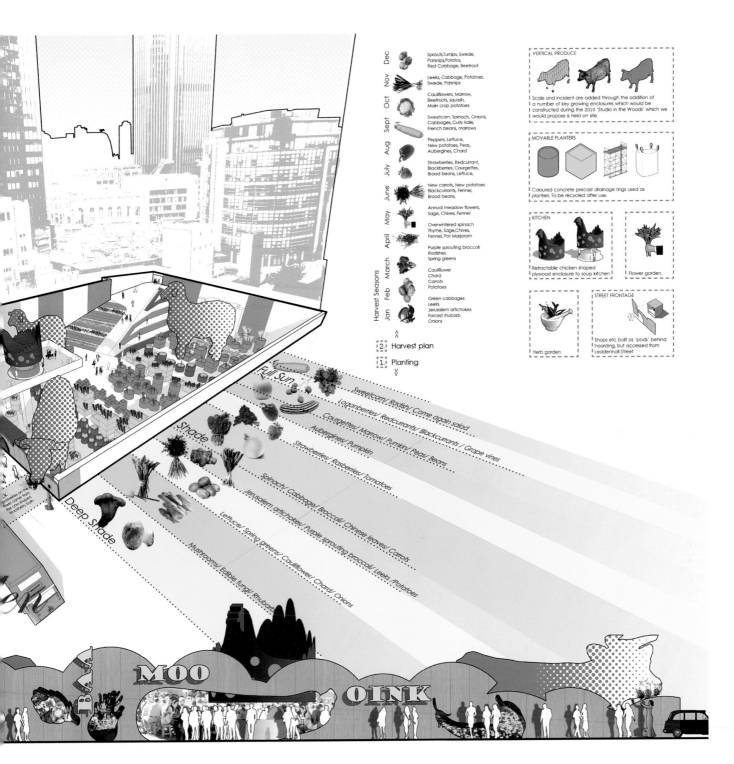

ABOVE: Proposed programming for the currently unused site.

PUBLIC FARM 1

WORK ARCHITECTURE COMPANY AND MOMA PS1 CONTEMPORARY ART CENTER
LONG ISLAND CITY, NEW YORK

Public Farm 1 (PF1) was an installation created by WORK Architecture Company (WORKac) during the summer of 2008. It was set in the courtyard of New York's PS1 Contemporary Art Center, part of the Museum of Modern Art (MoMA), and was part of the MoMA PS1 Young Architects Program that brings together new ideas in art, music, and architectural experimentation at the PS1 gallery in Long Island City, New York. PF1 exemplifies an artistic approach to local food provision; it is a sculptural intervention within a densely urban setting that illustrates how a bold, creative approach can celebrate densification and at the same time bring the systems and infrastructure that sustain a city from the periphery to its core.

PF1 was inspired by recent developments in organic and biodynamic farming and the potential to integrate these into urban environments. The aim was to demonstrate a self-sustaining, self-regulating, and multidimensional productive urban garden that could be integrated into a local community to create "a magical plot of rural delights inserted within the city grid."[36] Although PF1 was temporary, it suggests a potential future for many wasted and underused spaces within the city, including the estimated 14,000 acres of unshaded rooftop in New York City. During the exhibition, the PS1 café was provided with fresh, organically grown fruits and vegetables from the courtyard, reducing the food miles of the meals served. After the final harvest of the growing season, the temporary installation was composted, reused, and recycled.

Architecture, structure, materiality, and function were integrated at many scales using the city grid as an organizing pattern for the design. WORKac was keen to use a structural material that would be recyclable and biodegradable, hence the use of cardboard tubes normally reserved for construction formwork as the primary building material. The tube segments were preassembled in a cellular pattern resembling a honeycomb to create a field of forty "daisies," or groupings of seven tube segments organized into radially patterned units. Certain individual tubes extended to the ground as columns to support the entire structure, while the majority were kept short to create a large open canopy that shaded the area underneath.

The project required careful attention to technical detail. The tubes served as

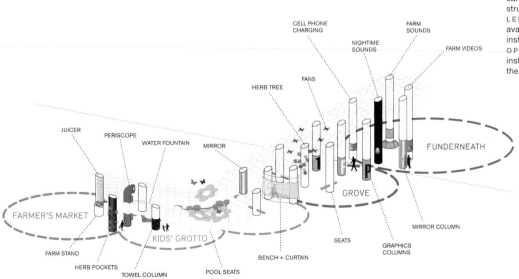

ABOVE: A close-up view of the various plants, drip irrigation system, and cardboard tubes that form the basic structure of the PF1 installation.
LEFT: The various programming available to visitors beneath the installation, on the ground level.
OPPOSITE: A side view of the installation in its final position in the PS1 courtyard.

FARMER'S MARKET

JUICER

PERISCOPE

WATER FOUNTAIN

MIRROR

HERB TREE

FANS

CELL PHONE CHARGING

NIGHTIME SOUNDS

FARM SOUNDS

FARM VIDEOS

FUNDERNEATH

GROVE

KIDS' GROTTO

FARM STAND

HERB POCKETS

TOWEL COLUMN

POOL SEATS

BENCH + CURTAIN

SEATS

GRAPHICS COLUMNS

MIRROR COLUMN

planters for herbs, fruits, and vegetables that were selected to thrive in the unique conditions created by PS1's hot summer courtyard. A diverse mix of twenty-three types and fifty-one varieties of plants were used, planted to bloom in succession throughout the summer. Each daisy was planted with a single species to create a changing field of color and texture over the whole installation. A daisy's center tube was left empty to allow PF1's farmers to climb into the structure and harvest the produce. In collaboration with elasticCo textile design studio, a special picking skirt was designed for ease of movement while the reaper harvested in midair. The plants were brought to PS1 from the greenhouses of the Queens County Farm Museum and the Horticultural Society of New York (HSNY) and transplanted into fabric containers called Smart Pots. The planters were designed to support the weight of plants grown in GaiaSoil—an ultralightweight growing medium made of recycled polystyrene foam, topped off with a layer of jute and two inches of compost. HSNY's Green Team, WORKac, and volunteer farmers maintained the system throughout the summer.

In addition to providing structural support, each cardboard column was designated to serve as a "zone" for a specific experience or interaction for visitors walking beneath the structure. These included a solar-powered juicer that could make fresh vegetable cocktails, a dried-herb column, a periscope that provided views of the fields above, a water-spouting column that recirculated the water from the pool, and a towel column to dry off. The Grove zone featured perpendicular tubes jutting out from main columns that provided seating and others that grew herbs; hanging fans wafted the herbs' scents down to the crowds. The Funderneath zone used video and sound to bring farm animal experiences to PF1.

The installation was self-sufficient in terms of water and electricity. A rainwater cistern collected more than 6,000 gallons of water from PS1's roofs over the summer, feeding a drip irrigation system that delivered a controlled amount of water to each planter. An array of eighteen photovoltaic panels generated electricity that was stored in ten lead-acid batteries and powered PF1's electrical equipment, including pumps that moved the rainwater throughout the farm and activated the fountain spout.

This project is one in a series of experiments by WORKac in "rurbanization," in which urban space is fused with rural characteristics, and the features of small-town life and cosmopolitanism coexist. These projects explore how to integrate urban density with productive areas, bringing together food production and consumption through mixing and layering of space usage in the city. WORKac now incorporates food dimensions explicitly in many of its design projects, regardless of scale or context, and maintains a web of relationships with the different specialists to whom they were introduced during the PF1 exhibit.

ABOVE: A view of the underside of the PF1 tubes. Fans blow air over the various herbs positioned on the arms that radiate from the main column to waft their scents down to the visitors below.
RIGHT: A cross-section of the cardboard tubes and their components; the structure was designed to hold a lightweight compound called GaiaSoil, first developed for green roofs.
OPPOSITE: Detail of crops grown and the innovative picking skirt that enabled staff to harvest produce comfortably while standing in the middle tube of each "daisy."

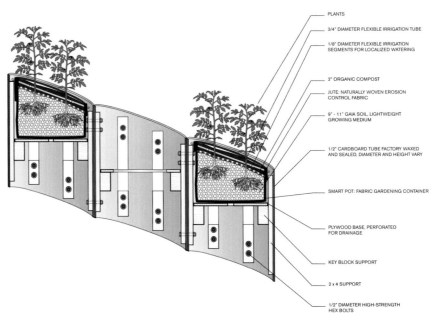

PLANTS

3/4" DIAMETER FLEXIBLE IRRIGATION TUBE

1/8" DIAMETER FLEXIBLE IRRIGATION SEGMENTS FOR LOCALIZED WATERING

2" ORGANIC COMPOST

JUTE: NATURALLY WOVEN EROSION CONTROL FABRIC

9" - 11" GAIA SOIL, LIGHTWEIGHT GROWING MEDIUM

1/2" CARDBOARD TUBE FACTORY WAXED AND SEALED, DIAMETER AND HEIGHT VARY

SMART POT: FABRIC GARDENING CONTAINER

PLYWOOD BASE, PERFORATED FOR DRAINAGE

KEY BLOCK SUPPORT

2 x 4 SUPPORT

1/2" DIAMETER HIGH-STRENGTH HEX BOLTS

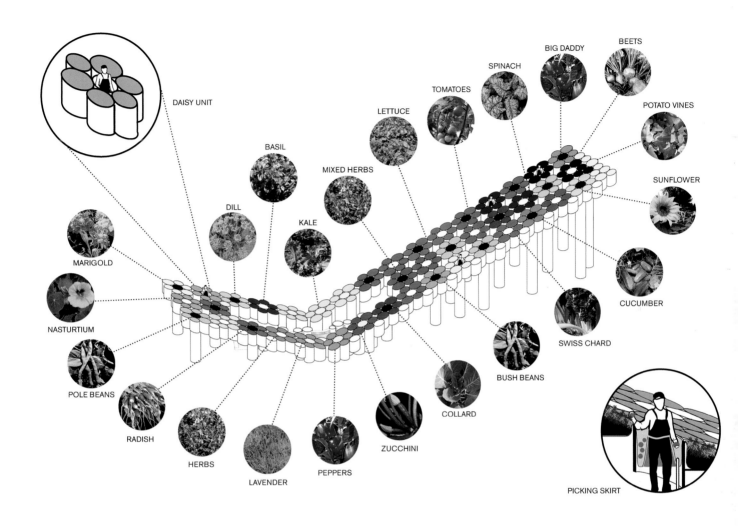

DAISY UNIT

MARIGOLD

NASTURTIUM

POLE BEANS

RADISH

HERBS

LAVENDER

PEPPERS

DILL

BASIL

KALE

MIXED HERBS

LETTUCE

ZUCCHINI

COLLARD

TOMATOES

SPINACH

BIG DADDY

BEETS

BUSH BEANS

SWISS CHARD

POTATO VINES

SUNFLOWER

CUCUMBER

PICKING SKIRT

1 FARM SOUNDS
2 FARM VIDEOS
3 MIRROR COLUMN
4 NIGHTSKY COLUMN
5 CELL PHONE CHARGING
6 GRAPHICS COLUMNS
7 MISTERS
8 SEATS
9 HERB TREE
10 BENCH + CURTAIN
11 POOL + SEATS
12 TOWEL COLUMN
13 PERISCOPE
14 WATER FOUNTAIN
15 HERB POCKETS
16 JUICER TABLE

LEFT: A longitudinal section of the installation showing both plants and programming underneath; the drawing gives an indication of scale in relation to human heights as well.
BELOW: The PS1 courtyard filled with visitors to the museum during a summer concert.

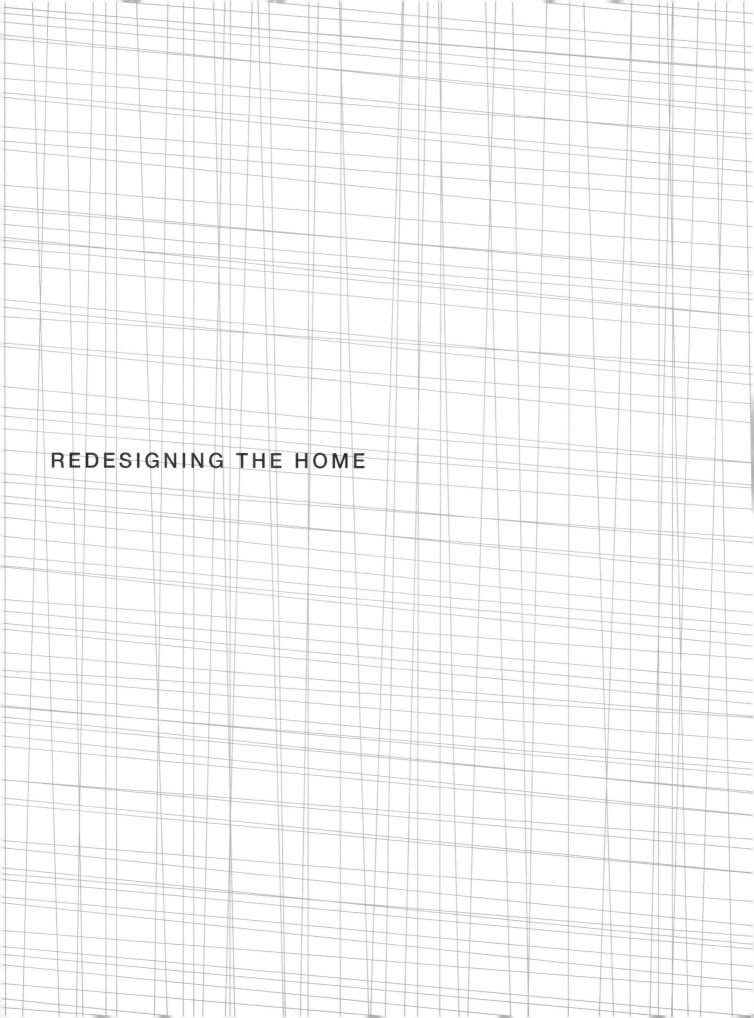

REDESIGNING THE HOME

In Western countries, until the advent of suburbia as a dominant pattern for human settlements, if a family's lot was large enough for a yard, there would be food growing there, which would be processed and later stored in the home. Over the course of the twentieth century, food production in the West moved gradually away from the home and dwellings became spaces solely for consumption. Architects, planners, and developers consequently created neighborhoods with little or no space devoted to productive gardening. Now, with more than half the world's population living in cities,[1] and poorer households spending more than half their income on food,[2] designers are beginning to reconsider this relationship. Reintroducing food production to housing promotes good health, reduces food miles, lowers the carbon footprint, and increases food security. Food production at home has emerged as a crucial part of sustainable design.

Food and the Home: An Intertwined History

In the Western tradition, gardens came in many forms. Carefully planned food-producing gardens were found in medieval monasteries, as seen in the plan for the famous example of the monastery of St. Gall from the ninth century. It includes chicken houses, vegetable plots, a combined orchard/cemetery with fourteen different types of fruit-bearing trees, and fountains.[3] These forms evolved into the sometimes highly stylized kitchen gardens, or *jardins potagers,* in France and elegant, geometrically designed herb knot gardens in Elizabethan England. Edible crops were also key to the self-sufficiency of secular homes from the Middle Ages on, and houses were built with features to aid in the processing, preservation, and storage of food. Significant space was devoted to larders, cold cellars, and icehouses.

Growing food at or near home continued to be essential to living well for many centuries. Witold Rybczynski notes in his book *Home: A Short History of an Idea* that in the seventeenth century, the Dutch prized "three things above all else: first their children, second their homes, and third their gardens."[4] In Italy during the Renaissance, the orangerie was devised in order to support citrus trees during the winter, and their popularity soon spread across Europe. Homes for royalty and the very rich often included them, and many, such as the grand structure in Kuskovo, Moscow, that dates to the 1760s, are replete with Italianate architectural detailing. In England and North America, especially in the Victorian and Edwardian eras, homes for the wealthier classes were designed with conservatories, which were similarly filled with edible and decorative plants.

In the early twentieth century, housing proposals by Raymond Unwin and Barry Parker in England for the garden suburbs inspired by Ebenezer Howard continued to incorporate food production and processing. This idyllic suburban landscape for the wealthy was adopted by the expanding middle and working classes, as expressed by Charlie Chaplin in his 1936 film *Modern Times*, where he can reach through the window to pick an apple and open the kitchen door to milk a passing cow. This all changed as the home went through a dramatic transformation centered on the modern kitchen. Ranges replaced hearths, and small pantries or cupboards were

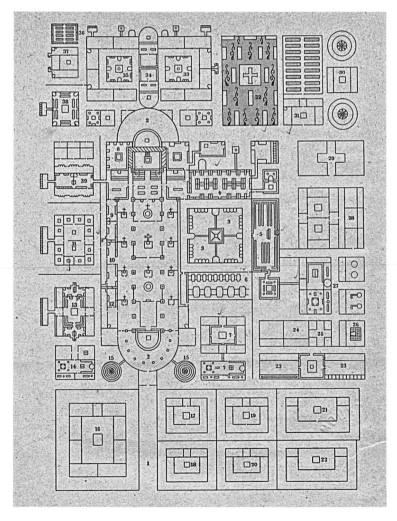

LEFT: Plan for the unbuilt, utopian monastery of St. Gall in Switzerland and its surrounding elaborate gardens, 819–826 C.E.

swapped for the large root cellars of the past. When the electric refrigerator/freezer eventually replaced the tiny icebox, storage of fresh—as opposed to dried or canned—foods became possible at home. On a larger scale, industrial freezers used by importers and distributors meant food could be shipped from where it was being grown in season to where it was in demand out of season—every kind of fruit and vegetable became available year-round. This made canning and drying unnecessary and is seen as one important catalyst that severed the connection between the home and garden. The practice of gardening had to be revived when shortages during the world wars made growing vegetables and herbs essential.

Pastures associated with manor houses and prosperous land owners were also a feature of the European landscape from the Middles Ages onward, and while they served as grazing land for sheep and other farm animals through the eighteenth century, the vast swaths of green became a symbol of wealth that eventually evolved into the suburban lawn in North America. Resource-devouring lawns with no functional purpose now compete with productive gardens for space, and, in fact, many municipalities still forbid the use of these spaces for grazing or vegetable gardening out of a desire to create homogeneous-looking neighborhoods, as do arbitrary homeowners association regulations.

The Edible Home

In response to the problem of potentially productive spaces going to waste, designers have begun to reclaim lawns and to repurpose ideas from historical gardens. Artist Fritz Haeg creates "Edible Estates" through his Regional Prototype Gardens (see page 132), which he presents as an attack on the supremacy of the manicured lawn. The growing recognition that with world population shifting to live in cities, housing patterns require new ideas about how food production may be incorporated within—or attached

to—urban residences. Strategies are emerging for mid- and high-rise apartment buildings, underused parking lots, and densely built blocks that contain little or no productive green space. Vancouver's Mole Hill Community Housing Project, for example, eliminated much of the lawn in a neighborhood of restored heritage homes and replaced it with numerous edible plants and community gardening plots as well as a pedestrian-friendly laneway that serves as a recreational space (see page 116).

New technologies and building types also provide unprecedented opportunities to include food production in and around housing. Romses Architects, in their Harvest Green proposals for Vancouver, rethink the North American obsession with the suburban lawn and address the potential of yard space. They suggest creating a community based on various prefabricated housing units united by a productive zone along rear laneways. This configuration provides densification for the city's existing grid and enables backyards to become a resource for fresh food, energy harvesting, rainwater collection, pocket parks, and more (see page 123). Hydroponics and other technologies are also being incorporated into exterior surfaces, sunspaces, and balconies, allowing occupants to harvest produce even without stepping foot outdoors. Increasing numbers of architectural competition entries, student projects, and conceptual designs include at least one feature related to food production, and many of these ideas are now being realized as built projects.

Conversely, in developing countries, innovative ways to reuse readily available, inexpensive materials—such as plastic beverage containers—as wall and window box gardens are being devised in order to supplement residents' diets. A report on this type of initiative to the Food and Agriculture Organization of the United Nations suggests that through providing low-cost, technologically appropriate containers as well as basic training in horticulture, food security can indeed be increased.[5] In the crowded, informal housing settlements of Brazilian *favelas*, the production of bananas, vegetables, and herbs in containers on rooftops, balconies, and next to paths has been encouraged as part of a nationwide zero hunger, or *fome zero*, campaign. A McGill University project in Colombo, Sri Lanka, similarly enabled food production in tiny alleys between small homes through the use of recycled containers often fastened directly onto house walls. Other components of buildings, such as water systems, heating systems, and organic wastes, can often also be given a second purpose, especially in dense cities and towns where a lack of access to sunlight and space is the main constraint on food production.

Appropriate scale has been found to be the critical design criterion in redesigning the home: at what scale do the different technologies and systems

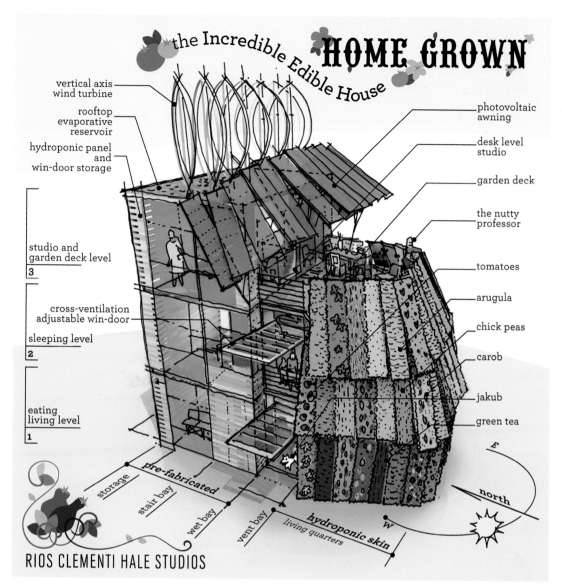

HOME GROWN

the Incredible Edible House

vertical axis wind turbine

rooftop evaporative reservoir

hydroponic panel and win-door storage

studio and garden deck level
3

cross-ventilation adjustable win-door

sleeping level
2

eating living level
1

photovoltaic awning

desk level studio

garden deck

the nutty professor

tomatoes

arugula

chick peas

carob

jakub

green tea

E

north

W

storage

pre-fabricated

stair bay

wet bay

vent bay

hydroponic skin
living quarters

RIOS CLEMENTI HALE STUDIOS

LEFT: Drawing by Rios Clementi Hale Studios in response to a 2009 *Wall Street Journal* invitation to design the most energy-efficient, self-sufficient house imaginable.
OPPOSITE: Facade of the Italianate orangerie designed in 1761 at the Kuskovo Estate in Moscow, Russia.

providing food, energy, water, and waste processing work together most effectively? In some cases, the same technologies proposed for citywide initiatives would help achieve efficiency in individual homes, making a simultaneous implementation of both the ideal situation.

Realized and Conceptual Projects

Many conceptual, even futuristic, proposals for productive housing forms have emerged in recent years, furthering the conversation and inspiring others. Happily some of even the most cutting-edge designs have been implemented. Analyses of one particular housing type and the opportunities for food production that it offers seem to spur the generation of the most innovative ideas.

One interesting example of a highly experimental form is the conceptual Incredible Edible House, designed by Rios Clementi Hale Studios. This unusual and eye-catching project was devised when the *Wall Street Journal* invited four leading architects to design the most energy-efficient houses they

could imagine without budget constraints or considerations of cultural norms. The designs were featured in a 2009 article entitled "The Green House of the Future."[6] Rios Clementi Hale's compact urban house designed for high-density neighborhoods enables residents to reduce their food miles to almost zero by creating a system in which food is planted on the front facade's articulated vertical surfaces through the use of a hydroponic skin that would support the growth of a variety of vegetables and fruits. The roof deck, which is also partly covered with wind turbines and a photovoltaic awning for both shade and energy production, acts as a rainwater collection system that in turn irrigates the productive spaces.

Today's architecture students, who have grown up with an awareness of environmental concerns, often incorporate food production into housing projects. The multifamily Vertical Farm Arcology project by Gordon Graff at the University of Waterloo[7] is inspired by the vertical farm proposals of Dickson Despommier. His residential complex includes programming for three mutually beneficial pursuits: wastewater filtration, hydroponic farming, and power generation via an anaerobic digester. Six and a half stories of the building are devoted to vertical, hydroponic food production—producing enough to feed 1,000 people—and the expansive roofs are slated for community gardens.

At a smaller scale, another unbuilt student project, the Edible Terrace, explores food production in a typical row house in the United Kingdom (see page 136). Every space on the lot and every built component of the house is reexamined for its food-growing, processing, or storage potential. Sustainable energy, water, and waste-processing systems are also fully integrated. This house may be too expensive to be practical for its likely middle-class occupant, but the ideas embedded in the proposal demonstrate ways of simultaneously resolving energy use and food production issues that support the goals of both.

On the other side of the globe, the Agro-Housing project (see page 140) is a competition-winning

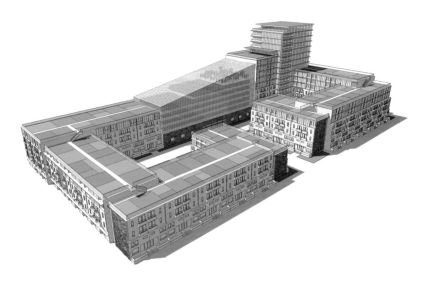

proposal due to be constructed in Wuhan, China, that looks at the needs and skills of Chinese citizens with an agrarian past who have recently relocated to dense urban areas. High-density apartments are mixed with facilities for food production through the incorporation of south-facing greenhouses with drip irrigation systems and trellises for climbing plants. The building contains rainwater collection systems and communal spaces to create a neighborhood atmosphere where residents can share expertise and practice the farming lifestyle they recently left behind.

The Center for Urban Agriculture is an unbuilt conceptual design for Seattle, Washington, by Mithun Architects that similarly demonstrates how a substantial redesign of the standard apartment building complex could change people's expectations of a residential structure and greatly reduce the amount of energy they use (see page 144). In this project, a significant portion of the structure—constructed

mainly of repurposed shipping containers—is given over to food production, and residents on every floor and in every unit are able to participate.

Examples of successful built structures that are currently inhabited and meld urban agriculture with affordable housing include Maison Productive House in Montreal (see page 126), 60 Richmond Street in Toronto (see page 148) and Curran House in San Francisco. Richmond Street is a recently completed project designed by Teeple Architects that incorporates strategies designers have proposed for much larger projects. The building was designed to provide affordable cooperative housing for hospitality industry workers. The residents' expertise in food service makes the large, productive garden that is integrated into the middle of the structure's deep block perfectly suited to its occupants. Large voids cut both horizontally and vertically into the mass of the building provide space for container and vertical gardens, as well as social space and additional light and air circulation for the units.

Curran House, a high-density affordable apartment building by David Baker and Partners Architects in San Francisco, houses over sixty low-income families. The roof is a key space for food production, and its border of citrus trees and grouping of individual container gardens made from galvanized livestock feed tubs provide individual plots while defining a common area where all residents interact and relax, a feature new to the city's low-income housing developments.

Currently under construction in the Melrose neighborhood of the South Bronx, New York, Via Verde, or Green Way, is another large-scale, affordable housing project organized around a series of gardens, some ornamental and others productive, that begin at street level and spiral upward through a series of programmed, south-facing roof gardens that culminate in a sky terrace. The plethora of green spaces incorporated into the structure provide residents a variety of options for physical activity right above their living quarters.

As interest in urban agriculture spreads and projects are realized, there will be a corresponding need for demonstrations to illustrate and educate participating residents about gardening in urban spaces. City Farmer, a nonprofit center in Vancouver, British Columbia (see page 120), has been filling this need and promoting sustainable practices since 1978. It promotes the creation of attractive as well as functional garden objects and offers hands-on, skill-building workshops on topics ranging from worm composting to organic gardening for the general public, schools, and other interested groups. City Farmer's two Web sites have also become an international clearinghouse for information about urban agriculture.

Conclusion

Admittedly, there are difficulties inherent in adapting existing structures perhaps not meant to bear heavy loads on rooftops—or access to them—to accommodate urban agriculture. The most difficult challenges to the initiatives that would convert lawns and waste spaces into productive gardens discussed in this chapter, however, are often the regulations imposed by individual neighborhoods, community associations, condo complexes, and towns. These are often very specific about what can be planted in the area covered, and require requesting special permission—which has no guarantee of being granted—to create a productive space. Other challenges are posed by the lack of knowledge among inexperienced gardeners about how to learn to plant and care for edible landscaping, particularly using organic means. A certain set of skills is required for success, and a community that embraces individual gardens and landscaping must also take responsibility for giving people the resources they need to keep productive areas healthy and tidy.

Lifestyle issues are also important considerations critical to the success of small-scale gardening. A perception exists that the time needed to maintain garden plots can be hard to carve from busy twenty-first-century lifestyles, for example, yet shopping expeditions often offset the time commitment needed for watering, weeding, and harvesting. Communal gardening and yard space rental to SPIN (Small-Plot Intensive) farmers are two solutions that lessen individual commitments, and others are available. Other challenges to adoption of food production at home include fears of soil contamination in urban cores. Ever-increasing urban density and traffic are problems that must be addressed through soil testing and improved practices, and which necessitate interventions by both governmental agencies and civic organizations.

The variety of projects in this chapter shows the many adaptations designers have already made to a host of housing conditions and challenges, including climate, sun angles, available surface areas, energy harvesting, and the need for social spaces, among other concerns. This is only the beginning of a rich dialogue between designers and inhabitants that has the potential to create residences that can accommodate a new way of life in which food production is an integral part of the home environment.

ABOVE: Rendering of the proposed Via Verde rooftop gardens, designed to bring green space and urban agriculture to depressed areas of the Bronx in New York City.
OPPOSITE: A rendering of the Vertical Farm Arcology proposal.

MOLE HILL COMMUNITY HOUSING
DIALOG, SEAN R. MCEWEN ASSOCIATED ARCHITECTS, DURANTE KREUK,
THE CITY OF VANCOUVER, BC HOUSING, AND MOLE HILL COMMUNITY HOUSING SOCIETY
VANCOUVER, BRITISH COLUMBIA, CANADA

The Mole Hill Community Housing Project is a model of neighbor-hood preservation and community food production. Twenty-six late-nineteenth-century Victorian homes in Vancouver's West End were restored and renovated to provide 170 units of affordable housing. The homes were almost destroyed during Vancouver's move to increase open space in the 1950s by razing large swaths of buildings, and this historic block was slated for demolition. Advocacy by organizations including the Mole Hill Living Heritage Society and the Friends of Mole Hill succeeded in preserving the structures. In the end, the city even invested in the project, moved buildings to empty lots on the site, and leased the land for sixty years to the Community Housing Society. In 1999 the transformation began, and the block became an "eco-friendly, diverse, affordable, heri-tage, socially-aware, and mixed-residential environment . . . a boon to all of the West End."[8] The project won several awards, and the historic buildings are now protected—original details were kept where possible to preserve their unique character, including window frames, doorframes, fireplaces, balustrades, and wooden flooring.

In addition to market-rate units, subsi-dized rental units, and a few privately owned buildings, six nonprofit organiza-tions have a presence on the site. One house is leased to the Coast Foundation Society to conduct a live-in program for youth that have mental health issues, temporary housing is provided by Heart House for recovering heart patients, and McLaren House and the Dr. Peter Center provide housing and services, respectively, for HIV-positive Vancouver residents. A YMCA day care center occupies part of one house, and the grounds are home to a farmers' market on Saturdays. Four cars owned by the Cooperative Auto Network are parked on the site, which also includes a secure bike room.

While the preservation and adaptation of the historic buildings was exemplary, the transformation of the spaces around the buildings may have been the most innovative aspect of this project. For example a storm water management sys-tem diverts rainwater into an ornamental pond rather than into the city's storm system. This pond is part of a garden cre-ated between two houses, creating a lush space that connects the main street to a lane that runs behind the houses.

Pathway over existing underground service lines to allow
for access to underground utilities.

New Development on vacant lot at Thurlow and Comox is
low scale and in character with houses on block.

Re-routing of Lane occurs only where existing surface
parking is located.

Utility poles to remain.

Existing sheds and garages to remain.

Lane allows for City, Service, Emergency Vehicle and
Resident access. Through traffic is discouraged.

Pedestrian-scale lighting.

Community Gardens and Composting.

Lane becomes Public Greenway and Parkland

LEFT: A perspective rendering of early suggestions for Mole Hill transformations.
ABOVE: An aerial view of the revitalized Mole Hill site.
OPPOSITE: Views of community garden plots and renovated houses.

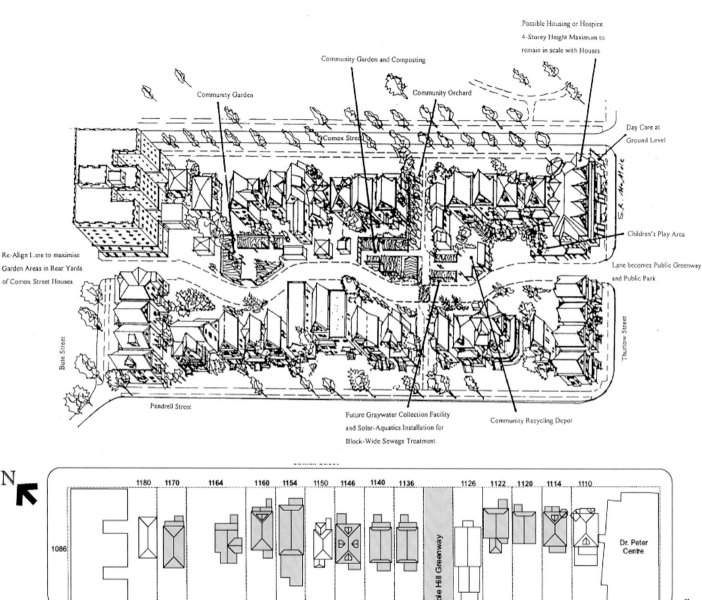

Possible Housing or Hospice
4-Storey Height Maximum to
remain in scale with Houses.

Community Garden and Composting

Community Orchard

Community Garden

Day Care at
Ground Level

Comox Street

S. A. Architect

Children's Play Area

Re-Align Lane to maximise
Garden Areas in Rear Yards
of Comox Street Houses.

Lane becomes Public Greenway
and Public Park.

Bute Street

Thurlow Street

Pendrell Street

Future Graywater Collection Facility
and Solar-Aquatics Installation for
Block-Wide Sewage Treatment.

Community Recycling Depot

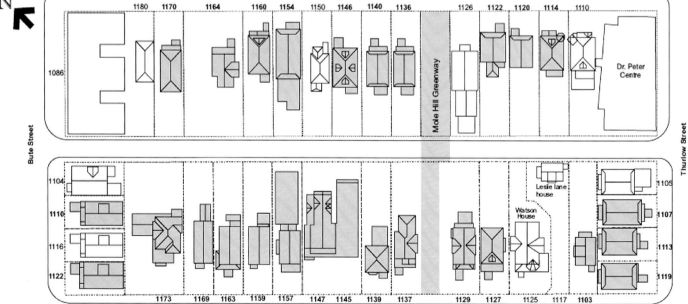

N

| 1180 | 1170 | | 1164 | | 1160 | 1154 | 1150 | 1146 | 1140 | 1136 | | 1126 | 1122 | 1120 | 1114 | 1110 |

1086

Mole Hill Greenway

Dr. Peter
Centre

Bute Street

Thurlow Street

1104 | | | | | | | | | | | | | | | 1105

1110 | | | | | | | | | | Leslie lane house | | | | 1107

1116 | | | | | | | | | Watson House | | | | 1113

1122 | | | | | | | | | | | | | | 1119

1173 | 1169 | 1163 | 1159 | 1157 | 1147 | 1145 | 1139 | 1137 | 1129 | 1127 | 1125 | 1117 | 1103

Pendrell Street

The transformation of the lane itself was truly remarkable; it became the heart of this community. Paving was narrowed from its original width of 33 feet to 20 feet (6.1 meters) during redevelopment to create space for edible and ornamental landscaping that creates a parklike atmosphere. This new pedestrian-priority path meanders through the block. At the intersection of the laneway and the garden/walkway, a square with benches and artwork beckons. Community garden plots on both sides of the lane cluster around it, and the garden view is enjoyed by residents relaxing on the benches. Plots are divided into wood-sided raised beds, making them tidy, well defined, and well drained. Additional community garden plots nearby are shared between Mole Hill tenants and other West End neighbors. This "living lane" has emerged as the true focal point of the Mole Hill community.

The Mole Hill gardens and pedestrian lane had been informally planned by community members long before they were realized. Architect Sean McEwen sketched a perspective of this idea some years before the garden was built, illustrating the Mole Hill community's aspirations for their lane as a multiuse, pedestrian-priority space. He recalls, "One interesting aspect about the community garden is that it was installed over a weekend during construction of the housing, in space that was earmarked for gardens only once the housing construction was completed. The garden was constructed by activist gardeners seeking to humanize a construction site. There are no plans or construction details as such. The gardens were laid out on a simple grid system, and constructed with donated and found materials." In an effort to strengthen the connection of Mole Hill to its surrounding community, about half the raised beds here were made available to neighboring residents.

Garage buildings that originally lined the lane have been replaced by a community laundry, workshop, and recycling area. Composting bins along it are used for both domestic refuse and green waste from all the new planted areas. As a result, this back lane that could have become an unsightly dumping ground or dangerous area has been transformed by food production. Besides adding an important function to these green areas, community gardening in the lane ensures the constant bustling presence of neighbors for safety, while avoiding the maintenance costs of caring for green areas. It is important to note that none of this transformation could have been realized, however, without a single owner, the city, controlling all buildings, lots, and the laneway.

"As contentious as the development of the block was at the time as an affordable housing project with a pedestrian priority lane incorporating community gardens," McEwen states, "ten years later, Mole Hill is an accepted and valued part of the Vancouver landscape."[9] Faculty and students from the University of British Columbia, members of Heritage Vancouver, and other groups regularly tour Mole Hill as a model of a successful urban agriculture program.[10] Mole Hill shows how sustainable planning for development can provide a sense of community, preserve architectural heritage, and create opportunities for growing food in the heart of a dense urban core.

TOP: The old laneway, transformed into a public greenway.
ABOVE: The curving pedestrian path as seen from the laneway.
OPPOSITE, ABOVE: The finalized community development plan showing the relationship of the pedestrian laneway to community gardens and housing.
OPPOSITE, BELOW: Plan of the original Mole Hill Community Housing site condition.

CITY FARMER
MICHAEL LEVENSTON
VANCOUVER, BRITISH COLUMBIA, CANADA

City Farmer is a nonprofit enterprise that was launched in Vancouver by pioneer urban agriculture advocate Michael Levenston. It promotes sustainable urban gardening via a demonstration garden in Kitsilano, one of the city's older neighborhoods. This small garden illustrates how to grow food in tight urban spaces, and within this space experts conduct training sessions on worm composting, backyard composting, organic gardening, natural yard care, and water-wise gardening for school groups, apartment dwellers, and the general public.

ABOVE, LEFT: The City Farmer house seen from the garden.
ABOVE, RIGHT: Fancifully decorated worm composting bins.
OPPOSITE: The City Farmer overall site plan and plant list.

City Farmer promotes economic and environmental sustainability, and espouses inclusiveness and simplicity. The result is a tranquil and beautiful garden. When Levenston started City Farmer over three decades ago, food production was absent from the suburban landscape and from the imagination of its residents. "It was a bit of an oxymoron," he recalls. "If rural, you're agricultural; if urban, you're a modern consumer. Our outlook at City Farmer was mostly aimed at the backyard farmers who had veggie gardens. It was a noncommercial demographic and we wanted to teach them about the environment through our organization."

The garden's beauty comes from the way the planting beds are combined with a series of interesting structures that rethink the aesthetics of standard garden elements. It includes several small buildings constructed with varying, unusual technologies. The most striking are a cob—clay, sand and straw—outdoor oven and a cob garden shed, both covered with a flowery green roof and sculpted to look Gaudíesque, with free-form floral and vegetal designs incorporated into clay plaster walls covered with litema, an African clay-coating technique that is decorative as well as protective. The storage shed whimsically suggests a gnome's home but houses equipment.[11] Sustainability was a prime consideration, so traditional, earthen materials were used for the shed's walls, and reclaimed concrete was used for its foundation.

Emphasis on craft and handwork is further reinforced by the garden's aesthethic embellishments. Davide Pan, a well-known metal sculptor, created the garden's striking gate from recycled materials. Visitors are greeted at the

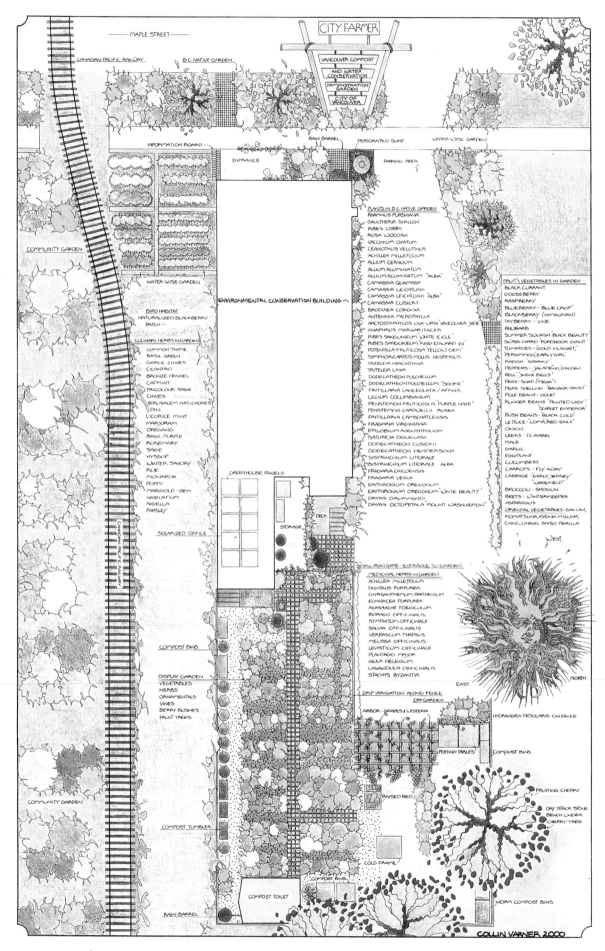

threshold by this artistic piece featuring reconfigured rusted garden tools welded together. A sizable wood compost bin is painted with lively, bright, giant worm graphics. These components add humor and fantasy to the garden, showing how creativity applied to a garden's ordinary, utilitarian components can add delight beyond what plants or birds themselves offer to a green space.

While at its core City Farmer is a local training center for urban agriculture and composting, its outreach has become international. Levenston's pioneering Web site has become a massive clearinghouse of information about urban agriculture events, initiatives, and knowledge from around the world.[12] Online slide shows teach gardening skills to remote visitors that are normally taught at City Farmer's on-site workshops. Lessons available through the City Farmer Web site demonstrate worm composting methods, how to build a green roof for a shed, how to implement natural lawn care, and how to design solutions for conserving, capturing, and storing water. These digital demonstrations complement the numerous physical examples that are scattered all around City Farmer's small yard and adjacent alley, such as porous grass paving grids, composting bins, a composting toilet, different types of rain barrels, and cold frames—low lean-to greenhouses—that enable an extended growing season. This small space and its vast Web site are rich sources for designers and gardeners working with urban garden plots, whether they are local to Vancouver or based far afield.

HARVEST GREEN
ROMSES ARCHITECTS
VANCOUVER, BRITISH COLUMBIA, CANADA

In 2009 the Architectural Institute of British Columbia and the City of Vancouver organized the FormShift ideas competition to generate new concepts on how to "improve the city's livability through greener, denser development."[13] The competition addressed the city's desire for carbon neutrality as expressed in the city's Climate Action Plan, the adoption of the 2030 Challenge,[14] and its EcoDensity Charter, which sets goals for compact development.

Harvest Green 02 was the winning entry to a competition category addressing small residential sites in established, lower density, Vancouver neighborhoods near public transit. This project is motivated by the same concerns about the North American obsession with the suburban lawn as the "Edible Estates" projects by Fritz Haeg, yet its solutions are fundamentally different. Romses Architects proposes allocating a rear portion of Vancouver's lots for a variety of productive new uses, making backyards

a resource for the city. Drawing inspiration from traditional Chinese Hutong mixed-use developments and newly approved Vancouver bylaws that create greater potential for laneway housing, Harvest Green 02 proposes the gradual introduction of a new web of food and local energy production that would overlay existing residential neighborhoods. The laneways become green energy and food conduits feeding and fueling the city in increasingly sustainable ways while subtly increasing urban density.

Romses Architects propose that a 33-foot (10-meter) zone at the rear of each lot along existing laneways be designated as a "flex zone" where a variety of private or community uses are permitted, including secondary residential rental units, community gardens, craft studios, small businesses, corner stores, shared car co-op parking, communal energy harvesting, rainwater collection, or even pocket parks. As these various uses take root, the designers propose that the standard 20-foot (6-meter) laneway gradually be reduced to 13 feet (4 meters). Permeable pavers, bioswale drainage features, and passing points for traffic moving in opposite directions would fill this area. The project promotes a steady transformation of laneways into

ABOVE: A neighborhood of modpod units; green
spaces are adjacent to, as well as integrated
directly on, residential structures.
PREVIOUS PAGE: A view down a productive
laneway lined with modpods.

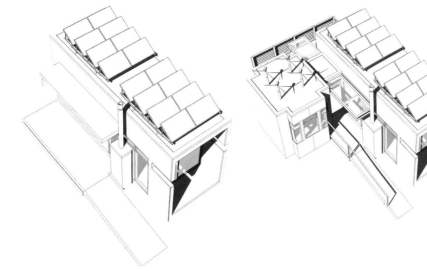
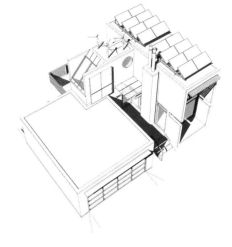

a tightly knit, pedestrian-friendly network of communities. These improve the resilience of the city and its ability to adapt to changes in trends and the needs of its inhabitants.

Prefabricated units outfitted with various energy, water collection, and food production technologies are key to this scheme. The modular laneway buildings—known as modpods—are highly utilitarian and compact, and come in a variety of sizes and configurations. Recognizing the constraints of many laneway sites, the system is designed to be flexible and adaptable to accommodate different solar orientations, and local constraints. Every site would be analyzed and paired with the custom modpod that best fit the needs of the site's potential for energy harvesting and food production. Photovoltaic arrays on rooftops and walls as well as trellises, living walls, and green roofs maximize collection on exposed surfaces.

The modules can be "stacked, mirrored, rotated vertically, and sited in multiple configurations and sizes to create a dynamic and varied built-form experience along the green street laneways."[15] As with many other prefab proposals, the modules fit on a standard flatbed truck and are designed for adaptability, ease of construction, and minimal construction waste. They also sit on four raised corner supports to reduce site disturbance. On their own the modules appear to be similar to other prefab housing proposals. When strung together to form a community, however, these units become a surprisingly flexible, organic, and adaptable neighborhood that intensifies Vancouver's laneway productive capacity. Food production can be carried out on rooftops and facades of the modpods, within and between them.

Given the renewed interest in intensification of existing urban areas and the need for new sources of energy and food production, this proposal is an interesting model for further exploration and has generated interest in Vancouver and in other locations, such as China. The project also highlights the need for partnerships between the local community, developers, and municipal authorities to realize innovative ideas. Since Harvest Green 02 does not require a grand master plan or large intervention, but addresses major issues through small-scale changes, implementation is viable and achievable. By utilizing existing infrastructure and space, taking a gentle and subtle approach, it is conceivable that this project could set in motion a gradual transformation of older, lower-density residential neighborhoods.

ABOVE AND BELOW: Five configurations of modpod units; the flexible system responds to different needs.

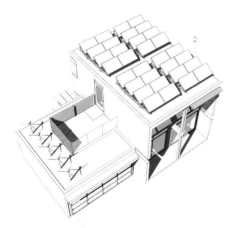
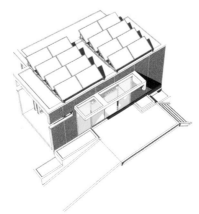

MAISON PRODUCTIVE HOUSE
PRODUKTIF STUDIO DE DESIGN AND DESIGN 1 HABITAT
MONTREAL, QUEBEC, CANADA

Maison Productive House (MpH) in Montreal, Canada, illustrates the potential for integrating food as a central theme in small, mixed-use, urban infill and renovation projects that address sustainable living. It also demonstrates the use of shared facilities, including for food production, in which residents take responsibility for their impact on the environment. The project is a labor of love for designer and urban agriculture researcher Rune Kongshaug,[16] who believes "good design integrates living, working, food production, transport and leisure."

Partly inspired by BedZed, the mixed-use exemplar sustainable project near London, the project was occupied in 2009, having had to overcome a number of financial, political, and legal hurdles in order to get underway. In the end, Kongshaug founded Produktif Studio de Design, bought the site himself, and acted as general contractor to help realize the project. The shared space on the site is managed by Design 1 Habitat (D1H), an innovative nonprofit organization that Kongshaug initiated to work with the construction industry to develop, monitor,

and manage building solutions that are "economically sustainable, socially just and ecologically sound." D1H works to find solutions to the effective implementation of urban agriculture in the context of residential development, recognizing the specific challenges for the mainstream North American market.

Location was an important consideration for MpH. The building is sited in the working-class Point St-Charles neighborhood near Montreal's historic Atwater Market and is easily accessible by public transit, bicycle trails and

pedestrian routes. The 15,700-square-foot (1,460-square-meter) project revives a condemned three-story building. The building itself forms an L-shape to create a protected microclimate for plants inside the lot. Five apartments are located behind the preserved facade, while three newly constructed townhouses form a wing perpendicular to the street, rising four stories above the small rear garden and orchard to provide 10,000 square feet (900 square meters) of total residential space. The 1500-square-foot (140-square-meter) townhouses are more compact and vertical than typical Montreal housing types. They contain three bedrooms and a large open-concept kitchen and living room opening to a loft space above, with a ground-floor room that can be sublet or used as an office. This creates a multifunctional design allowing the units to evolve with

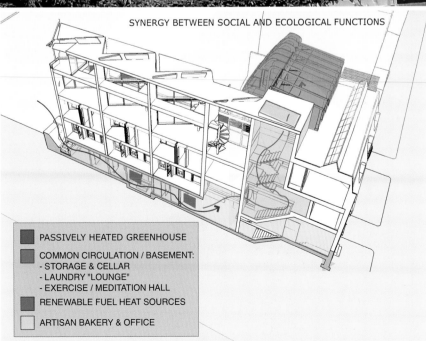

SYNERGY BETWEEN SOCIAL AND ECOLOGICAL FUNCTIONS

ABOVE: An L-shaped configuration formed by the site's original building at the rear and new townhouses at right shelters plants in the interior courtyard.
LEFT: Social and ecological functions are designed to support each other in the pursuit of creating a healthy, gratifying living space for all residents.
OPPOSITE: Windows on the new townhouses are angled to maximize solar exposure.

- PASSIVELY HEATED GREENHOUSE
- COMMON CIRCULATION / BASEMENT:
  - STORAGE & CELLAR
  - LAUNDRY "LOUNGE"
  - EXERCISE / MEDITATION HALL
- RENEWABLE FUEL HEAT SOURCES
- ARTISAN BAKERY & OFFICE

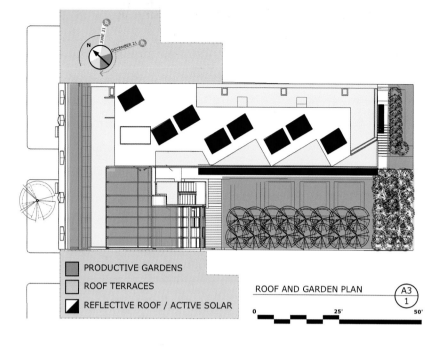

ROOF AND GARDEN PLAN $\binom{A3}{1}$

- PRODUCTIVE GARDENS
- ROOF TERRACES
- REFLECTIVE ROOF / ACTIVE SOLAR

0   25'   50'

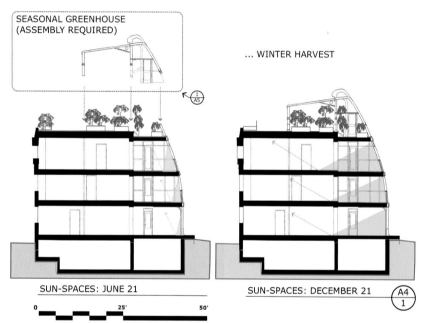

SEASONAL GREENHOUSE
(ASSEMBLY REQUIRED)

... WINTER HARVEST

$\binom{1}{A5}$

SUN-SPACES: JUNE 21

SUN-SPACES: DECEMBER 21 $\binom{A4}{1}$

0   25'   50'

TOP: Plan for the roof level of the apartments; nearly all the space is allocated to either energy collection or food production.
ABOVE: A seasonal greenhouse extends the growing season through the harsh Montreal winter.
RIGHT: A section view showing how photovoltaic panels on the roof are oriented to catch solar energy.
OPPOSITE, TOP: A dumb-waiter system allows residents access to heavy planters.
OPPOSITE, RIGHT: Planters on the roof, open for use in spring and summer.

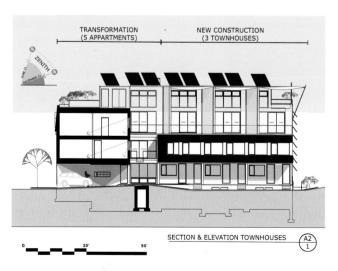

TRANSFORMATION (5 APPARTMENTS)   NEW CONSTRUCTION (3 TOWNHOUSES)

SECTION & ELEVATION TOWNHOUSES $\binom{A2}{1}$

0   25'   50'

the changing needs of occupants. The upper floors are rotated to maximize sun exposure, creating a sawtooth massing arrangement; living spaces open onto south-facing triangular terraces while loft spaces open onto north-facing terraces. Both provide space for growing. Some of the apartments have large, south-facing sunspaces, which serve as buffer zones against the harsh Montreal winter and form a continuous curving facade, which sweeps up the rear of the building to enclose a communal greenhouse.

On the ground floor, a bakery and a small design studio occupy about 1,100 square feet (100 square meters). Parking is minimal—two spaces in the former carriageway are provided for a carshare that is made available to the surrounding community as well as to MpH residents, who get a discount. Common spaces include a large rooftop garden, a four-season greenhouse, a sauna, exercise room, meditation room, and laundry facilities. Collectively, these occupy over 30 percent of the total building, or about 4,700 square feet (440 meters).

Each resident has an allotted space for growing in interior and exterior productive spaces. Individuals are responsible for kitchen gardens and sunspaces. Common spaces are managed and maintained by D1H, including the herb and berry gardens, fruit trees, and vertical gardens on landings of fire escapes, four-season greenhouse, and gardens on the shared roof. The fourth-floor greenhouse receives unobstructed sunlight throughout the day and uses two layers of low U-value polycarbonate glazing to trap the heat, extending the growing season. Waste heat from the building's sauna and bakery helps to warm the greenhouse, which has no other heating. A dumb-waiter system helps residents bring supplies up to the greenhouse.

Rainwater collection and gray-water recycling systems irrigate food-producing spaces. Rainwater is supplied to gardens growing salads and greens, and to balconies and sunspaces. Gray-water irrigation is used to supplement the lack of sufficient rain water. Timers coordinate the use of the collected water with the demand of the fifteen different gardens connected to drip irrigation systems. In total, 48 percent of the site is covered by food production, habitat, or other permeable surface, and much of the rest collects rainwater. Kongshaug is also looking for opportunities to connect to existing alternative distribution networks, including trading with community-supported agriculture schemes, for the supply of wood, plants, and livestock that cannot be produced on-site. The bakery has been used to explore relationships with periurban or rural flour mills and farmers.

Kongshaug predicts that MpH can achieve 60 to 80 percent energy autonomy; he would like it to produce more than it consumes in the future. The project is Novaclimat certified for energy efficiency, aimed to achieve platinum certification under the LEED system, and claims to be a Zero Emission Development. It features a broad spectrum of energy-saving technologies, including solar panels for the production of hot water, geothermal heating, Energy Star and low-water usage appliances, and radiant floors. Spaces are optimized for passive heating and cooling and many of the south-facing windows are equipped with exterior blinds. None of the spaces is air-conditioned, and residents are billed for individual energy consumption, which reduces wasteful practices. The building helps to minimize the use of materials containing volatile organic compounds and maximize the use of recycled materials.

This project demands a paradigm shift in North American urban living, design, and construction practices by promoting a radical vision of relative autonomy that aims to halve the carbon footprint of residents. However, not all residents are willing to accept such a challenge to their lifestyle, and this raises the question of whether similar projects would be embraced in the larger North American context. Cultural changes may be necessary before a wide-scale implementation would be successful. Even here, where residents are environmentally minded, a significant challenge occurred with acceptance of a common laundry, which is perceived as a feature of low-quality housing—each buyer was ultimately offered the option to sacrifice a shower for a private washing machine. The residents also underestimated the efforts required by gardening, so more productive space had to be taken over by D1H.

MpH is an important attempt to develop creative solutions for food production in an urban residential context. Furthermore the productive spaces of the MpH do more than simply produce food; they serve an important aesthetic and social role in the building. The aim is for MpH to encourage greener, healthier, and more fulfilling lifestyles. The project is intended to be a replicable and scalable solution that can be implemented in many other locations around the world, creating urban environments that nourish themselves and grow without continuing to degrade the planet.

TOP: Planters thriving in the enclosed greenhouse.
ABOVE: The ground-floor bakery.

ABOVE: Two views of the mature garden spaces.

EDIBLE ESTATES REGIONAL PROTOTYPE GARDENS
FRITZ HAEG
VARIOUS LOCATIONS IN THE UNITED STATES AND ENGLAND

Artist Fritz Haeg could be said to use the suburban lawn and its aesthetics as his medium; his *Edible Estates* installations challenge the underlying values and aesthetic inclinations that drive millions to plant vast expanses of water-hungry, pesticide-dependent swaths of green. Through his series of "Regional Prototype Gardens" planted across the U.S. and Europe, he contrasts the traditional domestic front lawn with a lush, aesthetically pleasing, climate-appropriate, edible landscape. Haeg planted the first garden in the series in 2005, and it continues to grow in scope and ambition. The gardens are purposely located in places calculated to serve as a vivid contrasts to their surroundings—surprise and shock are anticipated reactions and are intended to generate interest and curiosity about the transformations. As food author and activist Michael Pollan summarizes, "Lawns . . . are a symptom of, and a metaphor for, our skewed relationship to the land."[17]

Salina Art Center in Kansas City commissioned the first installation.[18] Haeg's design included two large circles symmetrically placed in front of a house, the first a low seating area and the other a high mound covered with herbs and tomatoes. He planted other perennials—strawberries, thyme, blackberries, and horseradish—in the yard to provide easy-to-care-for edible landscaping. One

of the homeowners, a plant geneticist, contributed perennial plants currently being developed as future grain crops by the Land Institute, where he works.[19] Strawberries covered the sloping ground that leads toward the street and grape vines were planted against the house. The garden was aesthetically striking and highly productive. Gardens in Lakewood, California, and Maplewood, New Jersey,

have also been installed. In Lakewood, Haeg replaced the front yard with two intertwining spirals, each containing a set of tall poles for climbing plants, such as beans, cucumbers, and melons. In Maplewood, the plan contrasted a gently curving path and high trellises with raised-bed planters arranged in a rational, rectilinear grid of square planters made from recycled plastic.

After these three suburban U.S. installations, Haeg created an urban garden commissioned by Tate Modern in London as part of their "Global Cities" exhibition in 2007, an edible garden at a council housing estate in Southwark, London, near the museum. This project was planned to be viewed primarily from above, as residents of the multistory apartment building gazed out from windows for a splendid aerial view of the communal garden. Portions of the lawn were removed for the introduction of curved beds; scores of volunteers added soil, mulch, and plants. Haeg explains his design: "The intricate design for the garden was inspired by

ABOVE: A vision of the Baltimore garden when fully mature.

LEFT: Curving beds in the Southwark, London, council housing garden recall the parterres of elaborate royal gardens.

OPPOSITE, LEFT: The Baltimore residence before planting, with its typical suburban lawn.

OPPOSITE, RIGHT: The Baltimore garden planted with numerous productive, edible plants.

the ornate, curvy, raised flower beds that you find in front of Buckingham Palace and Kensington Palace . . . The arabesque shapes allow for easy access to all planting beds and create two oval gathering spaces." Fruit trees create vertical interest and groupings of low-growing vegetables and herbs define the edges of the beds. This small bit of grass was redesigned as a useful community space that increases access to healthy food while offering learning opportunities for residents, particularly children, who delight in the garden.

Haeg's 2008 Austin, Texas, project commissioned by Arthouse was installed at Sierra Ridge, an apartment complex owned by Foundation Communities, a group that serves low-income families. Haeg created oval spaces for composting, rainwater harvesting, and gatherings. Walkways were created from thick layers of mulch, half-log benches were installed, and cypress logs and stone create raised beds filled with huge, hardy plants. These provide a welcome contrast to the utilitarian, uninspiring buildings they border. The next project, the *Descanso Public Demonstration Garden* in La Cañada Flintridge, California, was instead located in a park tended by professional gardeners who donated the produce to a local food bank. A redwood-frame structure in the shape of a classic small wooden house sits in the garden's center. The house's "front yard" is divided in half; the left side is a plain lawn, while the right overflows with an abundance of vegetables, berries, and edible flowers.

The Contemporary Museum in Baltimore, Maryland, commissioned Haeg's next project for the 2008 exhibit "Cottage Industry." This project moved back to the residential yard, and included circular mounded plantings to enable drainage, making a vivid contrast to the manicured lawns nearby. Following that project was a recent installation in New York City, the *Lenape Edible Estate* located at the Elliot-Chelsea House and The Hudson Guild Community Center in the Chelsea neighborhood of Manhattan.[20] Although it is a working edible landscape, it also demonstrates the possibilities for urban gardens in general and references the Lenape people's history of growing food on the island of Manhattan 400 years ago.

Haeg's work raises issues about the amount of manual labor needed to care for public edible landscapes. Haeg acknowledges that zoning and homeowners association regulations can prohibit residential areas from installing non-lawn landscaping. His thought-provoking projects reached out to a variety of audiences, however, challenging the wisdom of these established rules in numerous ways with an ultimate aim of changing them and getting potentially productive spaces under cultivation. His installations also question the relationship, or lack thereof, that contemporary urban and suburban residents have with the land in the neighborhood in which they live. Fritz Haeg shows that there is no dichotomy between the beautiful yard and the edible landscape; they can be one and the same.

ABOVE: The Descanso Public Demonstration Garden in California; the site is planted half with lawn and half with edible plants to show production potential of even a small plot.

ABOVE: The Lenape Edible Estate
in Manhattan before planting, top,
during planting by the residents,
above, and when mature, top right.

EDIBLE TERRACE
ANTHONY CAMPBELL AND JAMES WEST
BOLTON, UNITED KINGDOM

The Edible Terrace project is a reinterpretation of the English terraced house to meet the predicted needs of twenty-first-century urban living. Anthony Campbell and James West, two students from the Manchester School of Architecture, propose a self-sustaining living environment with organic food production, water collection, and renewable energy generation in a one-family house located in the post-industrial town of Bolton, in the north of England.

In some disadvantaged parts of the U.K., food and energy costs consume 60 to 90 percent of household spending. By reducing the need to spend so much on food and energy, the project simultaneously addresses social and economic disadvantages widespread in the area. At the same time, the availability of organic fresh fruit and vegetables grown on-site, with high nutritional quality, can improve the poor dietary habits prevalent in this community. The project thereby also proposes ways of reducing the impact of housing on climate change by generating food, as well as energy and water, on site.

The house form itself is simple. The compact terrace form reduces heat loss from the building and maximizes solar gain. The large south-facing, double-skin, glazed facade controls the internal environment. It acts as a buffer zone between the inside and outside while collecting energy and allowing light to penetrate the interior. The house includes a massive concrete core that acts as a heat store for solar gains and modulates the internal climate. Other energy-saving strategies include a highly insulated north-facing envelope and air-tight construction. A 190-square-foot (17.7-square-meter) array of photovoltaic panels on the roof consisting of thin film photovoltaic strips integrated into the roof glazing provides electricity and also allows natural light to filter down into the living space below. A vertical-axis wind turbine located in the backyard produces additional energy.

The project offers several different opportunities for food production that are symbiotically integrated with each other and the building's energy systems. A vegetable patch at the front of the property, planters on the rear terrace, and fruit trees in the rear yard provide fresh fruit and vegetables. Herbs can also be grown on the north-facing green roof, which encourages local biodiversity and provides insulation for the house beneath. Mushroom cultivation is proposed for often-shaded areas on the building's north side. Outdoor areas are supplemented by growing spaces between the two layers of glazing on the south facade, extending the growing season in this wet and cool climate into the winter months. Fruit and vegetable scraps can be fed to organically reared chickens and pigs that are kept in a 225-square-foot (21-square-meter) pen located in the rear garden. Manure from the animals provides feed for the tilapia fish that are proposed in the 10,000 gallon (44,300 liter) water tank

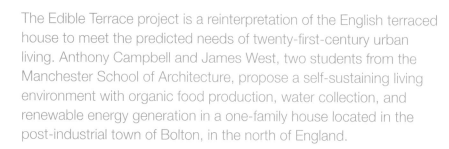

LEFT: The many components of the Edible Terrace are designed to be incorporated into one single-family, terraced house.
BELOW: A section through the house, revealing the different environmental and productive systems at work.
OPPOSITE: Backyards are devoted to food production; livestock living in the pen could eat scraps from human consumption, tilapia from the basement tank, even fruit that falls from the tree. The roof is also planted with herbs.

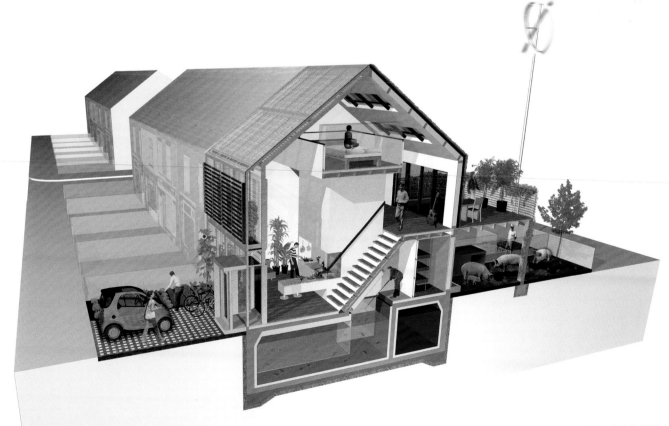

in the basement. In addition to being home for the fish, the water in the highly insulated tank acts as an interseasonal energy store heated by evacuating solar tubes on the south facade and provides heat directly to the under-floor heating system. The tilapia require temperatures between 73 and 82 degrees Fahrenheit (23 to 28 degrees Celsius), so they are restricted to the lower, cooler part of the storage tank. The fish are used as food for consumption by humans or fed to the pigs. Excess animal manure and human sewage is fed directly into a 600-cubic-foot (17-square-meter) biodigester in the basement, which is kept at an optimum temperature for anaerobic digestion of manure, to produce methane gas for cooking, and potentially for fueling a smart car. The waste from the biodigester can also be used as fertilizer.

Edible Terrace highlights the opportunities for rethinking standard housing solutions, and suggests many ways of integrating food systems with energy and water systems even at the scale of a standard terraced house footprint. This project provides a living environment and resources for two to three inhabitants on a typical terraced house lot in the area. It also addresses the social and economic impacts of living in "food desert" areas, solving the dilemma with its self-sustaining symbiotic organic food production. The concept has also been expanded for a larger unit on two combined lots, which would be suitable for four to six people. However, the project also raises questions of intent—should effort be directed at improving individual houses, or should appropriate solutions to energy and food supply come at the community or neighborhood scale? It is perhaps overly optimistic in its embrace of proposed technologies that may be more economically and technically realistic at the community scale. Recognizing this issue, Campbell and West also proposed a larger-scale community housing project, the Mancunian Wedge proposal, based on the same principles used in the Edible Terrace applied at a larger scale.

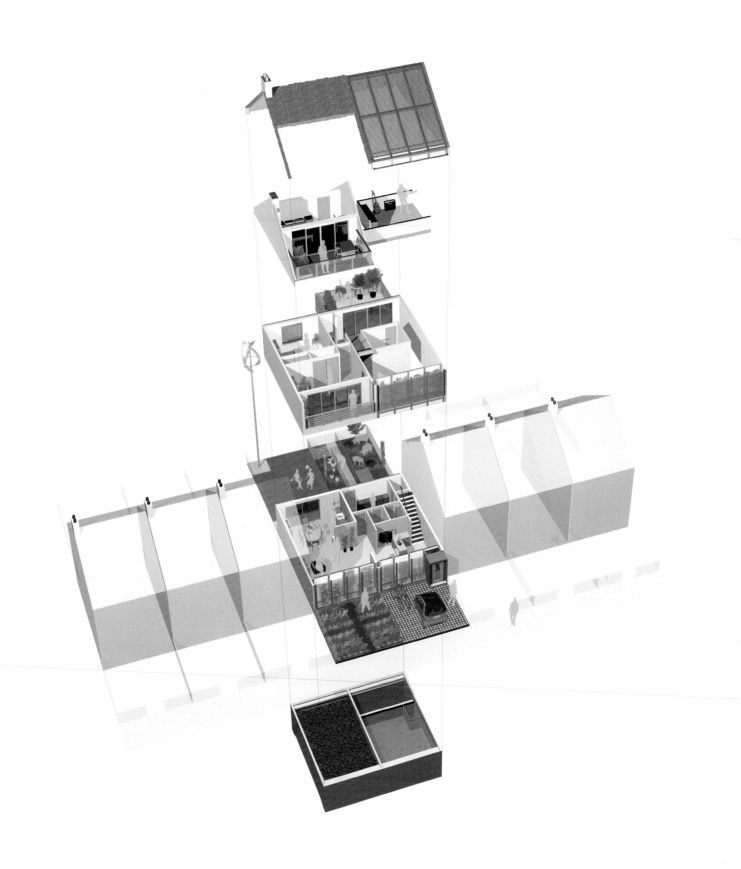

AGRO-HOUSING
KNAFO KLIMOR ARCHITECTS
WUHAN, CHINA

This proposal by Israeli firm Knafo Klimor Architects was the winning submission for a housing development in China designed to integrate food production, for the second Living Steel international design competition in 2007. Living Steel is a collaborative program managed by the World Steel Association aimed at stimulating responsible and innovative housing and construction solutions.[21] Agro-Housing, a multistory apartment block with a range of eco-friendly technologies and a variety of food-production spaces, specifically addresses housing needs in China, and has since been selected for realization in the city of Wuhan.

With previously unheard-of numbers of rural dwellers moving into cities to search for jobs created by China's dramatic economic transformation, many of China's megacities are facing enormous housing, resource, cultural, and planning problems. Rapid urbanization is putting tremendous strain on natural resources including food, water, and energy, and is exhausting infrastructure and transportation systems. Providing dwellings for millions of new inhabitants is only the first step in accommodating new residents, however. Acclimating families and individuals who have only known rural lifestyles to adapt to the anonymity,

pace, and various strains of urban life while making use of their rural skills and culture is the second. Putting these skill sets to use, particularly farming skills, would simultaneously close the gap in food security and make newly relocated residents feel they had become useful members of their new society.

The architects saw the Living Steel competition as an opportunity to explore ways of meeting the needs of an urban Chinese population with deeply ingrained rural values and a degree of independence. Agro-Housing proposes to integrate communal facilities, green technologies, and food production into

a twelve-story, 110,000-square-foot (10,000 square meter), 120-unit apartment building—a model for high-density living in China. Space is provided for residents to grow food of their own choosing, either for themselves or for sale to the surrounding community. This creates opportunities for families to supplement their income as well as diet. The aim is to strengthen community interaction while preserving knowledge. Agro-Housing was designed for low- to medium-income residents, although the concept can be adapted for a wider demographic.

The main productive space consists of five levels of double-story, 16-by-100-foot (5-by-30-meter) greenhouses incorporated into the south-facing facade to provide a total of 8,000 square feet (750 square meters) of growing space suitable for cultivating vegetables, fruits, flowers, and herbs. Designed for flexibility, the greenhouses can be partitioned into parcels of about 107 square feet (10 square meters) for individual families or configured as larger common gardens

where neighbors could farm together. Private areas can also be integrated within common areas to create access for nonresidents. Additional small, north-facing semiprivate gardens are interspersed among the apartments, creating neighborhood gardens within the larger building. As well as being productive, the gardens are an attractive amenity of the type not previously included in high-density housing.

Greenhouses are placed to enable light and sun to reach plants growing on trellises. They are designed to use a soilless media growing system created from coconut peat, mineral wool, volcanic ash, or perlite growing media. Drip irrigation is provided from rainwater captured and stored on the building's roof and gray water stored and purified within the building. The greenhouse is equipped with natural ventilation and is separated from the residential component of the building by a large vertical void that serves as a thermal chimney, drawing warm air out of the building through operable vents at its top and serves as a thermal buffer zone.

Agro-Housing provides a variety of community and semiprivate spaces

ABOVE: A rendering of the view from the street; the project's small footprint leaves ample room for bicycle parking and productive gardens at ground level.
BELOW: Small neighborhood garden spaces, terracotta cladding, and windows create a mosaic on the building's north facade.
OPPOSITE: Two-story greenhouses that comprise the bulk of the building's productive space face south to maximize solar exposure.

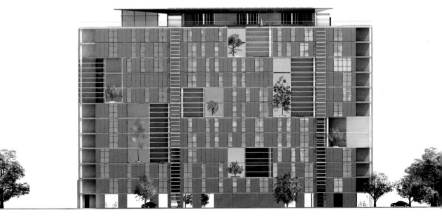

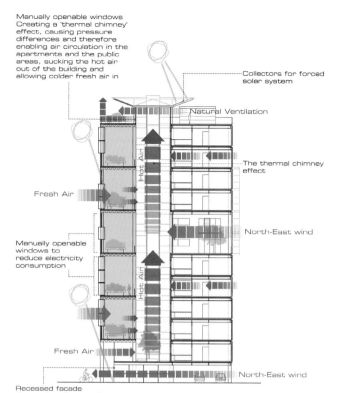

Manually openable windows Creating a 'thermal chimney' effect, causing pressure differences and therefore enabling air circulation in the apartments and the public areas, sucking the hot air out of the building and allowing colder fresh air in

Collectors for forced solar system

Natural Ventilation

Hot Air

The thermal chimney effect

Fresh Air

Manually openable windows to reduce electricity consumption

North-East wind

Hot Air

Fresh Air

Recessed facade decreasing direct radiation during the hot months

North-East wind

Summer

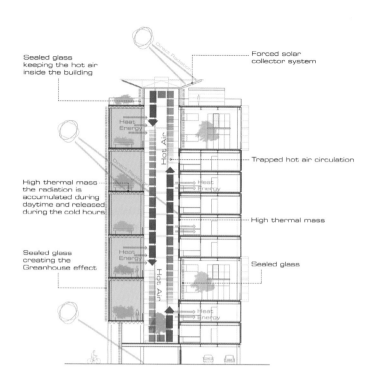

Sealed glass keeping the hot air inside the building

Forced solar collector system

Heat Energy

Trapped hot air circulation

Hot Air

High thermal mass the radiation is accumulated during daytime and released during the cold hours

Heat Energy

High thermal mass

Heat Energy

Sealed glass creating the Greenhouse effect

Sealed glass

Hot Air

Heat Energy

Winter

TOP: Sections showing the building's thermal circulation and natural ventilation system in different seasons.
ABOVE: The seventh-floor plan, showing how green spaces are distributed in relation to living spaces; the main greenhouse area runs along the bottom, or south, perimeter, and the void at the center denotes the thermal chimney.
OPPOSITE, TOP: An interior view of a double-height garden adjacent to a duplex apartment.
OPPOSITE, CENTER: The greenhouse as seen from the interior; specially designed beds hold soilless growing medium.
OPPOSITE, BELOW: A view from a double-height private garden to the common greenhouse space.

intended to create a sense of place and provide amenity within a high-density residential block. The building's small footprint frees up land for gardening and rainwater collection at ground level. First-floor space is reserved for bicycle parking and for a kindergarten—allowing children to be cared for close to home. The roof provides open green space for additional food production as well as space for a club for residents to host social gatherings and celebrations.

The project embraces many other green building practices to reduce its environmental impact, including recycled construction materials, gray water recycling systems, passive solar energy use, natural ventilation, and a ground source heat pump system. The structure's materials were selected for easy erection, dismantling, and recycling as well as economy. The load-bearing structure is steel while interior partitions are made of a customizable system of lightweight gypsum walls that allow residents to adapt their spaces and transform their apartments depending on individual needs. Working spaces can also be integrated within the apartments if needed. The building is clad in glass and terra cotta. Movable sunshades protect windows and interiors. Bathroom units are prefabricated complete with piping and ductwork.

The food production capacity of the Agro-Housing project will not likely meet the complete needs of its residents, but could provide a significant economic buffer while fostering rural cultural values in a dense urban environment. Skills and experience imported from the countryside can directly contribute to making cities more resilient to future changes in the availability of resources such as food, water, and energy. Agro-Housing proposes a future approach to housing that goes beyond typical sustainable housing strategies.

CENTER FOR URBAN AGRICULTURE
MITHUN ARCHITECTS, PLANNERS, AND DESIGNERS
SEATTLE, WASHINGTON

■ Baseline Biodiversity   ■ Living Building Biodiversity

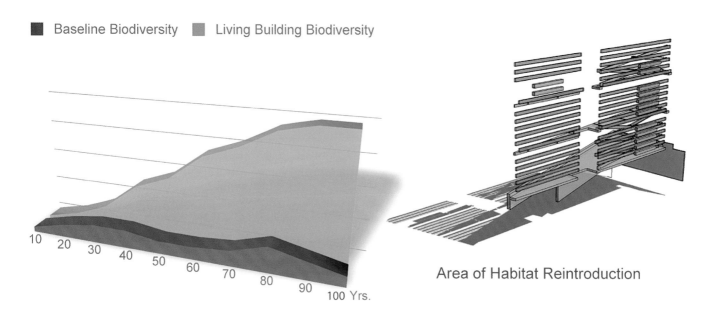

10 20 30 40 50 60 70 80 90 100 Yrs.

Area of Habitat Reintroduction

The Center for Urban Agriculture (CUA) proposal, designed by Seattle– and San Francisco–based Mithun Architects, Planners, and Designers, represents a rethinking of how we provide basic needs—shelter, energy, water, and food—within a dense urban building location. The proposal goes beyond including established green building practices for energy efficiency, water conservation, and use of environmentally sensitive materials to explore the notion of self-sufficiency and fundamentally reconsider how urban living can be resourced. It recognizes the impact of food supply on the carbon footprint of urban dwellers and seeks to provide alternative ways to supply resources locally.

The CUA won "Best in Show" award in the 2007 Living Future Competition organized by the Cascadia Green Building Council,[22] which challenged the construction industry to develop proposals that meet the very demanding requirements of the Living Building Challenge.[23] Unlike the LEED green building rating, Living Buildings have just sixteen demanding prerequisites that the building must meet.[24] They are based on the premise that it is possible to create buildings that function using the surrounding resources, collect their own energy and water, produce zero emissions, and use renewable resources. While there are not plans to build this actual project, several other Living Buildings are under construction, and Mithun maintains that a project like CUA is achievable using mostly off-the-shelf components.

Designed for an awkward triangular, 0.72 acre (0.3 hectare) site in downtown Seattle, the 250-foot-tall (76-meter-tall) building with a five-story base allows its location and lot to define its form, deriving its other formal characteristics from its orientation to the sun and its flexible, utilitarian structure. The first five floors of the building consist of a podium containing a restaurant and educational laboratory space filled with a large planted surface and chicken farm. Above the base, are eighteen stories of residential units made from the standard 8-by-8-by-40-foot (2.4-by-2.4-by-12-meter) recycled shipping containers (Seattle maintains one of the largest shipping container surpluses in the world), retrofitted as apartments off-site to speed construction and stacked within the superstructure using crane erection techniques.

The containers are arranged onto three "shelves," each of which holds stacks six containers high. These are oriented along a single-loaded, exterior circulation system that provides ideal natural ventilation and lighting while eliminating the need for heated public space. The 318 studio, one-, and two-bedroom apartments form a monolithic south-facing elevation with movable shading devices for the windows. These shades are clad largely in 34,000 square feet (3,150 square meters) of photovoltaic panels estimated to generate

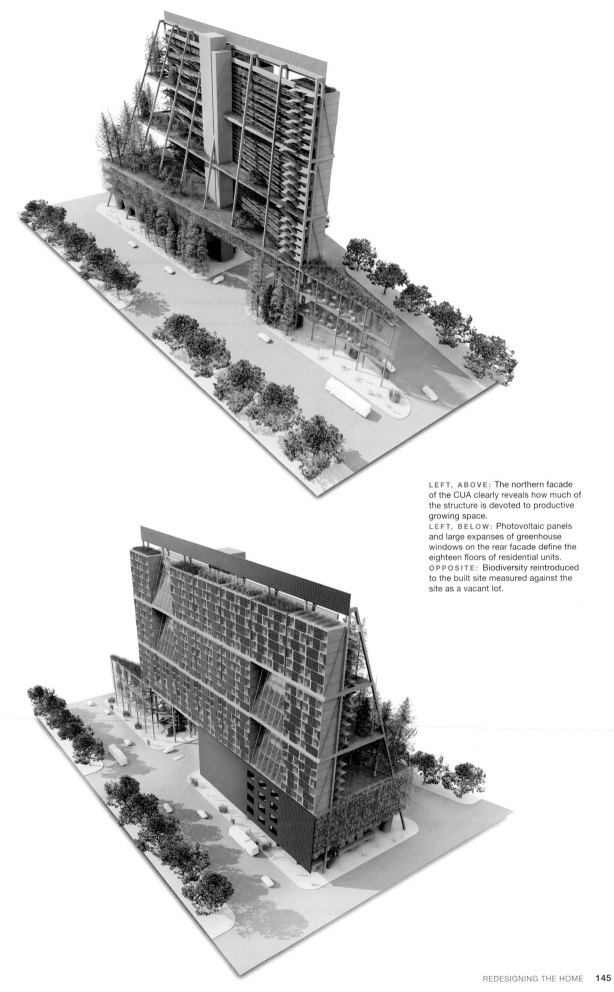

LEFT, ABOVE: The northern facade of the CUA clearly reveals how much of the structure is devoted to productive growing space.

LEFT, BELOW: Photovoltaic panels and large expanses of greenhouse windows on the rear facade define the eighteen floors of residential units.

OPPOSITE: Biodiversity reintroduced to the built site measured against the site as a vacant lot.

Embodied C from Baseline Food Production
Embodied C from Local Sources (<150 miles)

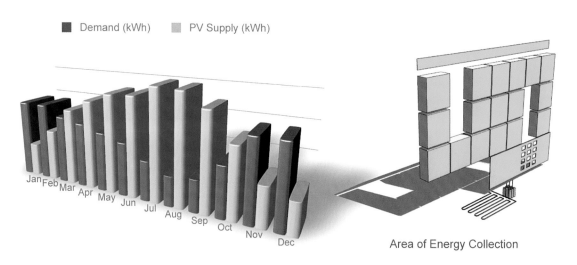

Metric Tons Carbon

Area of Food Production

Demand (kWh)   PV Supply (kWh)

Jan Feb Mar Apr May Jun Jul Aug Sep Oct Nov Dec

Area of Energy Collection

approximately 550,000 kWh of energy every year. This is stored in underground hydrogen tanks and provides approximately 90 percent of the building's electricity needs, with hydrogen fuel cells supplementing when required. A proposal linking the CUA into the surrounding urban fabric would utilize a large strip of south-facing land along the I-5 freeway corridor for further PV panels and electricity production.

Rainwater and gray water are collected and processed for use in the residences and agricultural operations, addressing the objective of water self-sufficiency. Forty-five large storage tanks holding 16.5 million gallons sit at street level and form a mazelike area for pedestrians to explore and enjoy. This storage capacity can also accommodate runoff from surrounding buildings and roads, acting as a reservoir.

The CUA exceeds Living Building requirements in its focus on making a substantial contribution to local food production, which is integrated with the other green technologies and systems. Three triangular terraces to the north of the residential units provide space for open-air tree and plant nurseries, fruit and vegetable cultivation, other types of native flora, a gathering space, and a 19,000-square-foot (1,750-square-meter) chicken farm. A number of greenhouses puncture the southern

facade, breaking up the residential block. A rooftop garden takes advantage of the sun's exposure and a restaurant and café in the base of the building provide the opportunity for nonresidents to sample food grown on the central site. The various terraces and rooftops add up to 1.35 acres of productive surface, which is impressive given the building's relatively small footprint. The designers do not identify whether the land is to be gardened individually by residents or in some collective way, but they suggest that the food grown in the building could be sold to the community and to the in-house restaurant below.[25] While industry estimates vary as to how much farmland is required to feed an individual, Mithun believes that by intensively cultivating the 185 square feet (17 square meters) available per resident, a substantial portion of their fruit and vegetable needs could be met. The designers estimate that this would save 2,000 metric tons of carbon dioxide from entering the atmosphere from food miles alone—a number equivalent to the typical annual electrical use of 273 households.

Another distinctive aspect of Mithun's proposal is its attention to boosting local biodiversity. By saving some of the rooftop garden and terrace space for what the designers call "patches" of native plantings, they intend to create habitat for a large number of native bird and insect species, an island of habitat within an urban context. Although this project was not conceived to play a role in a Continuous Productive Urban Landscape, it is indicative of the type of structure that could fit seamlessly into one if and when one was developed for Seattle.

Mithun employed a multidisciplinary team to conceive the design of the CUA, including landscape architects, architects, and even an ecologist to select an appropriate range of native trees and other plants that would simulate local native habitats. By promoting healthy ecologies in this way, the project not only fosters a social environment open to new urban patterns of living that facilitate local food production, but also points to a new type of physical urban environment in which agriculture projects can be integrated directly into the function of buildings and their surrounding spaces.

60 RICHMOND STREET EAST HOUSING CO-OPERATIVE
TEEPLE ARCHITECTS AND THE TORONTO COMMUNITY HOUSING CORPORATION
TORONTO, ONTARIO, CANADA

The Toronto Community Housing Corporation commissioned 60 Richmond Street East, an eleven-story, eighty-five-unit apartment building—the first cooperative housing project built in the city in twenty years. The project integrates numerous innovative strategies, including the incorporation of urban agriculture and food preparation spaces within the building's very core. Fifty-nine units in the complex were designated to replace units from the nearby Regent Park, a 1960s-era housing project. From inception, this midrise building was intended as affordable housing, primarily for hospitality industry workers. Design was not sacrificed and the project, by Teeple Architects, ultimately received the Best New Residential Building Award in Toronto, and the Ontario Association of Architects Design Excellence award. Despite the cost constraints, the project incorporated cutting-edge sustainable strategies, achieving a LEED Gold rating.

A productive garden on the sixth floor of the building—intended to supply some of the food for a ground-floor restaurant meant to be run by the residents— doubles as a community space while at the same time supplying fresh herbs and vegetables, demonstrating that productive gardens can also be social spaces and the goals of one reinforce the other.

Other design features include a top-floor cistern for the capture and storage of storm water runoff and an extensive green roof that helps to reduce the building's heat-island effect while insulating it. Setbacks also create private balconies and linear gardens that provide additional food-growing opportunities. Compost from the restaurant can also be introduced into the system as nutrition for the vegetation.

The sixth-floor garden was created by cutting large voids both horizontally and vertically into the mass of the building. These spaces, along with highly reflective, bright white exterior cladding on setbacks and cutouts that reflect the sun, introduce daylight into the center of the building for plants, and reduce the need for artificial lighting in the apartments. This form also contributes to natural ventilation throughout. A metal framework suspended from the sheltered eastern wall of the void accommodates a large vertical growing wall for further edible landscaping.

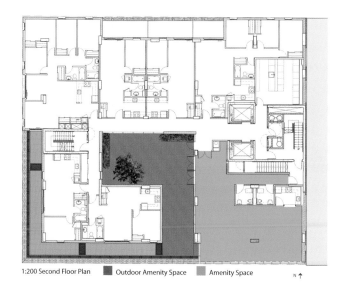

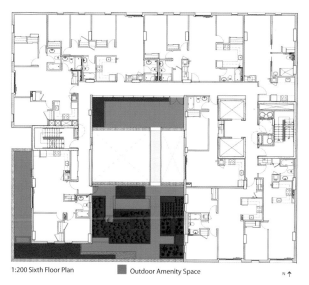

1:200 Second Floor Plan ■ Outdoor Amenity Space ■ Amenity Space   N ↑

1:200 Sixth Floor Plan ■ Outdoor Amenity Space   N ↑

TOP: The building's second-floor plan, left, and sixth-floor plan, right, showing garden areas and amenity spaces carved into the building's interior.
ABOVE: The building's extensive green roof insulates the building.
OPPOSITE: An early concept image for the project emphasizes its dedication to providing fresh fruit and vegetables for residents.

The white cladding of the recessed and void surfaces is sharply contrasted by charcoal-colored cladding on the outward-facing elevations. These dark facades also emphasize the playful yellows and oranges of the balconies. This rain-screen cladding is well insulated and designed so thermal bridging, a significant source of heat loss in many buildings, is reduced. Windows are carefully sized to balance the needs of daylight and ventilation with the increased heat loss that occurs even through the high-performance glazing and frames. Variations in mass, color, and setback depth create an overall balanced and rhythmic composition, providing additional visual interest.

Other design strategies include reuse of the foundation from the previous building on the site as shoring, and moves to encourage cycling, walking, and public transit by the residents. The Toronto streetcar runs near the site, and there is also a secure bicycle parking room on the ground floor—residents will not need to squeeze bicycles into the elevator or store them in their apartments. The parking garage contains intentionally few spaces, but an allocation was made for a car-share vehicle.

A collaborative design process incorporated ideas from an energy-modeling consultant, future tenants, the engineers, the client, and the architectural team. Although well thought out, it remains to be seen whether the growing spaces within the volume actually receive enough light to successfully grow vegetables in the raised beds. North-facing apartments overlooking a narrow laneway also receive reduced light, one intractable issue of designing in a dense urban location. Nevertheless, the project is exciting in its exploration of urban form that integrates food-growing spaces with other green building features. It proposes a possible future for urbanism that takes seriously the future availability—or lack—of resources including food, energy, and water, and embraces a living, growing form that responds equally to the city, a site's environmental conditions, and the sustainability imperatives of the future. It is a positive sign for the future of affordable housing that such building initiatives can receive funding and recognition—this shows how new technologies and strategies can contribute to the making of a sustainable, productive city.

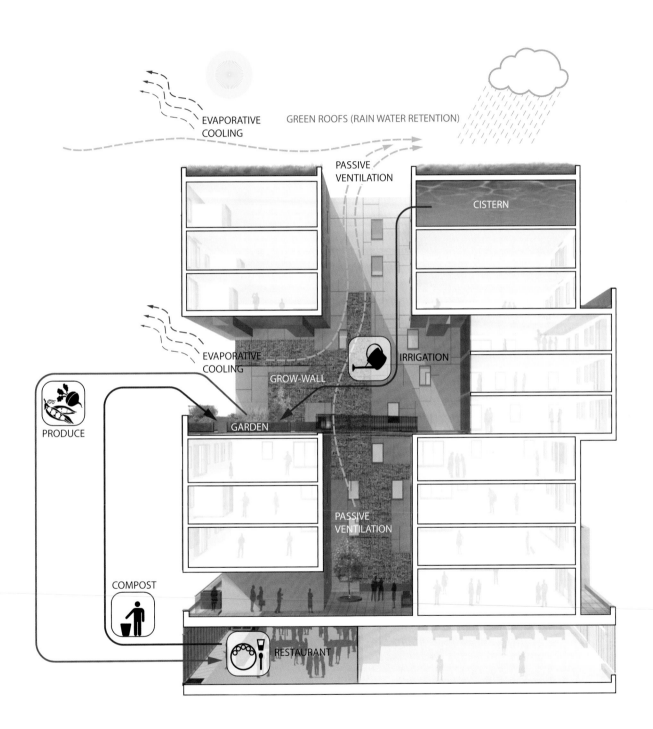

EVAPORATIVE
COOLING

GREEN ROOFS (RAIN WATER RETENTION)

PASSIVE
VENTILATION

CISTERN

EVAPORATIVE
COOLING

IRRIGATION

GROW-WALL

PRODUCE

GARDEN

PASSIVE
VENTILATION

COMPOST

RESTAURANT

PRODUCING ON THE ROOF

Urban development inevitably causes pressures on the availability and cost of land. Large, dense, and economically vibrant cities have a shortage of green space at ground level, but contain acres of unused horizontal space in the form of roofs, balconies, and terraces. Rooftop gardens can overcome the problems of high land costs and competition for land uses at street level while contributing to a city's overall health and climate.

Contemporary urban roofs are largely wasted spaces. Their surface area is underused and they are rarely considered as a resource. Instead they are covered as economically as possible and left untended until leaks make maintenance necessary. As cities become increasingly populous, however, the trend has been to harness their potential to generate energy, collect and store water, grow food, and provide relaxation spaces. The same factors that make conventional roofing systems difficult to maintain—such as unimpeded exposure to sun and precipitation—are beneficial to agricultural pursuits. Recent development of new solar technologies and increasing public awareness about the advantages of green roofs have made them increasingly in demand for a variety of productive uses—some building owners now even lease roof space to commercial farming organizations, others to energy companies. These new uses reduce a city's ecological footprint, and are generally recognized by green building rating systems such as LEED and BREEAM.

Environmental benefits for the local community include significantly reduced rainwater runoff, which reduces the stress on the municipal water systems. Rooftop gardens can retain 70 to 100 percent of the precipitation that falls on them in the summer and half that much in the winter. Green roofs also reduce the urban heat-island effect where solid, dark surfaces absorb the sun's rays and raise the temperature of the surroundings. Vegetation absorbs sunlight while reducing the warming effect and shading roofs. This creates microclimates of more comfortable temperatures in summer and lowers mechanical cooling needs in buildings. Furthermore during the winter months the soil contributes to insulating the building. Research suggests that plants on the roof can also act as a very effective filter of pollutants, removing as much as 95 percent of heavy metals such as cadmium, copper, and lead from runoff, and improve air quality by trapping and absorbing nitrous oxides, volatile organic compounds, and airborne particulate matter. Rooftop gardens in dense urban environments also create secluded, relaxing, natural spaces separated from the noise and commotion of urban activities below that are still easily accessible to city residents.[1]

The vast majority of green roof installations to date and most developments of new green roof technologies have focused on nonproductive green roofs. These often rely on thin soil layers and sedum or grass as vegetation. It is only recently that productive green roofs have begun to be seriously considered as an urban food production strategy. Planted roofs reclaim and transform unused or sterile urban spaces into productive green enclaves that contribute to urban ecology, health, and well-being, and rooftop gardening is a subset of urban agriculture with its own set of particular characteristics; it plays a critical role in the broader discussion of how to render cities productive.

Challenges
Growing food on a roof presents certain challenges. Protecting the plants from severe weather and windier conditions than exist at ground level, and collecting and storing enough rainwater for irrigation throughout summer months are major concerns.

RIGHT: Active rooftop gardening at Curran House in San Francisco, left, and an aerial view of the building's rooftop garden space, right.

Particular technical demands also exist, and many of the case studies in this chapter address one or more of them in an innovative way, including:

Load: Often the limiting factor in rooftop projects is the structural capacity of the roof to take the additional weight of soil. One cubic foot of wet earth weighs approximately 100 pounds (45 kilograms). These can be accommodated when planned for in new construction, but in existing buildings, load issues can be significant barriers to production. Placing heavier planters above the load-bearing columns or walls can help, but there are often other implications.

Soil: The weight and type of soil used in green roofs impact maintenance significantly. Delivery and replenishment of conventional garden soils can be a problem due to often-limited access points to the roof and the soil's weight. Most green roofs use some form of prepared, light-weight soil medium for this reason. Soil must be treated to ensure adequate nutrients for long periods of time, because it can be difficult to rotate or replenish.

Exposure: Rooftops have good exposure to sunlight, which is of course essential to plant growth, but can be exposed to high winds and/or snow drifts around vertical features, increasing loads. In some climates, hot and dry summertime winds can be very damaging or drying to both plants and their support structures, such as roof sheds and trellises, so shelter needs to be incorporated into the design. Vegetation needs to be carefully selected based on its ability to adapt to the microclimate of the roof.

Water and Irrigation: Relatively thin soil layers used on green roofs intended for horticulture must be designed with water retention and availability in mind. In some locations, rainwater collection and storage systems with irrigation systems may be required if city water is to be avoided. Locations with high summer rainfall, such as Tokyo or Shanghai, can often support rooftop food production using rainwater alone, but locations such as Vancouver, British Columbia, where only 24 percent of yearly rainfall occurs during the April to October growing season, significant irrigation may be required.[2] Specially designed planters can hold extra water to compensate for drying conditions.[3]

Waterproofing: Green roof technologies have advanced significantly in recent years to provide a range of options for waterproofing systems. On productive roofs, it is important to protect the waterproofing layer from potential tears caused by tools, machinery, and feet. Access to repair any leaks is necessary, and indeed an entire industry of leak-detection systems has emerged as the popularity of green roofs has grown.

Pollution: Vegetables grown in the city are often perceived as unsafe to eat due to concerns about pollution. However, studies suggest that vegetables grown on rooftops contain equal or lower amounts of heavy metals than many vegetables from the countryside. All leafy vegetables should be washed before consumption. It has also been shown that high levels of carbon dioxide present in urban areas can be beneficial to plant growth.[4]

Access and Safety: If significant amounts of food production are planned for a rooftop, access for deliveries of materials and removal of produce need to be planned, and as with any accessible roof, safety must be addressed. Most cities have specific requirements for fencing, depending on the level and type of access that is expected; this can lead to prohibitive costs.

Affordability: While rooftop planters or containers can be as simple as relatively inexpensive children's wading pools and are therefore affordable for anyone with access to an appropriate roof area, more sophisticated rooftop gardens can be expensive. Building code and insurance requirements and hiring an engineer can also add unexpected costs to the venture.

The Emergence of Productive Rooftop Gardens
Plants have been grown on roofs for many centuries, but the emergence of planned urban farming on roofs and balconies in the West has occurred only in the last few decades. Montreal's 1976 Rooftop Wastelands project on a community center rooftop was one of the first programs to examine the feasibility of small-scale, individually run food production plots within a Canadian city. This federally funded, 1,000-square-meter demonstration project was also used as a living classroom for an organic gardening course open to the public.[5]

Germany, Switzerland, and Japan have become known as leaders in the green roof movement by implementing incentives and code requirements promoting green roofs. The city of Stuttgart, Germany, passed a municipal bylaw in 1989 requiring all flat-roofed industrial buildings to install green roofs, a measure that was quickly adopted by other municipalities. In Switzerland regulations require that new construction must re-create any displaced green space, and existing buildings are expected to convert 20 percent of their roof space into a green area.

In warm countries, houses are commonly built with flat concrete roofs, which can provide growing space for vegetables—the climate allows for year-round crop production, so using them for this purpose has long been practiced and many community groups provide training and resources. In Russia, the first experimental gardens appeared on roofs of public buildings and apartment houses in St. Petersburg in 1993, following the dissolution of the U.S.S.R., and their popularity grew. Now municipal codes encourage their proliferation.[6]

In North America, cities including Toronto have passed local bylaws requiring green roofs—although not productive green roofs specifically—on all new, larger-scale commercial and residential construction. This has become recognized as a way of reclaiming the area of land displaced by the building. Innovative social housing projects such as Curran House in San Francisco include growing areas for residents. Condominium building developers are becoming aware that rooftop gardening opportunities can attract tenants. In Vancouver, British Columbia, the Freesia, a 181-unit high-rise, features sixty wood-frame raised beds, a tool shed, and garden lockers. Some plots are farmed by a local, independent urban agriculture business; others are made available to residents for a small fee. Successful projects like this stimulated, in turn, guidelines for inclusion of edible landscapes in the South East False Creek development (see page 44), and for the city of Vancouver as a whole.

Although small commercial rooftop farms, like Annex Organics in Toronto, existed in the 1990s, several recent larger-scale initiatives demonstrate the financial feasibility of rooftop food production. In New York, Brooklyn Grange and Eagle Street rooftop farms (see page 158) provide seasonal vegetables to the community and nearby restaurants. Both were initiated by urban gardening enthusiasts who wished to expand onto other New York roofs. SOLEfoods, based in downtown Vancouver, takes community involvement one step further—part of their mission is to provide employment to local residents so they may benefit from the economic stimulus generated by the business.

A key technology in the success of many rooftop gardens are hydroponic systems, which have been found to increase yields significantly. Three of the world's first commercial rooftop gardens, slated to open in 2011, are banking on them, including Forest House (see page 164). Gotham Greens in Brooklyn is building a 16,000-square-foot (1,500-square-meter) hydroponic farm powered by a 60-kilowatt solar-voltaic array on a roof. It expects to produce 40 tons of food annually, 70 percent of which will be sold to Whole Foods stores.[7] In Montreal, Lufa Farms, located on the roof of a two-story, 31,000-square-foot (10,300-suare-meter) office building near that city's central market, plans to plant fragile—but often tastier—varieties of vegetables for local consumers and restaurants that cannot survive shipment to distant destinations.[8]

Hotels, restaurants, and shops have also seen the benefit of producing food on their roofs for sale or use in their own kitchens. Toronto's Carrot Green Roof is exemplary in this regard (see page 170). Eli Zabar's Vinegar Factory Markets Rooftop Farm in New York City has been producing food commercially for sale in the market directly below on its 20,000-square-foot (1,850-square-meter) roof since 1995 (see page 164). A proposal for a concept grocery store and restaurant that combines hydroponic, aeroponic and aquaponic farming, named Agropolis, would give shoppers the unique experience of harvesting their own produce right in the store, where it would be growing in a soilless medium. Nutrients gathered from fish byproducts in aquaculture tanks below shoppers' feet would feed the plants in a closed loop ecosystem.[9]

Seattle Urban Farm Company grows vegetables in a 4,500–square-foot (420-square-meter) garden on top of the Bastille Café and Bar. Its menu features the "Rooftop Salad," whose ingredients are grown in glass-covered, raised beds and in inexpensive children's round, plastic wading pools.[10] Noble Rot restaurant in Portland, Oregon, also used thirty-nine wading pools on its 2,100-square-foot (195-square-meter) roof to grow staples for its menu. Each 45-inch diameter pool provided about 12 square feet of growing space. Fairmont Hotels provide a good example of initiative being taken at the corporate level; the chain converted roofs of several hotels for herb and vegetable production (see page 188).

A 2005 Green Roof Grant Program provided by the City of Chicago resulted in over twenty green roofs across the city. Several businesses took advantage of this unique opportunity, including Carnivale, Browntrout, the Uncommon Ground Restaurant (see page 168), and True Nature Foods. The latter is an organic market cooperative that created a low-budget 960-square-foot (90-square-meter) rooftop garden. Built by nonprofit organization Urban Habitat Chicago, it is a good example of a low-intensity, simple, cost-effective, loose-laid system. The

ABOVE: Vegetables growing on the roof of True Nature Foods in Chicago, one of the earliest retail ventures to grow the produce it sells on-site.
OPPOSITE: Trent University in Ontario, Canada, boasts an extensive rooftop garden; students harvest produce and use it in preparing food for sale at an on-campus, student-run café.

growing structure was placed on top of the existing roof membrane, which can accommodate the load because soil depths are kept to 3 to 4 inches. It grows a variety of herbs and vegetables—some not commonly found in supermarkets, like buckwheat, burdock, and comfrey.

Some of the lessons from this low-budget project have been the problems with access to the roof, which is only by a ladder, and the need for a proper staging area for managing the produce. Without this, bringing supplies, including the regular addition of organic fertilizer, and removing the harvest is difficult.

Rooftops are also increasingly used for educational projects. A well-established rooftop vegetable garden at Trent University in Peterborough, Ontario, has provided research facilities since 1994. The 9,000-square-foot (830-square-meter) garden, located on the flat roof of the Environmental Science Building, is used as a learning space for students in the Food and Agriculture Emphasis Program at the school and for research by Dr. Tom Hutchinson. Much of the produce goes to the Seasoned Spoon Café, a student-run, independent cooperative on the

university campus, which has a mandate to source local ingredients. Its plants surround ventilation stacks, benefitting from their warm exhaust air and shade. The garden provides temperature moderation for the spaces below in return, reducing the need for air conditioning in summer and providing insulation in winter. One lesson learned from this roof relates to access—until recently supplies had to be brought up through a boardroom below because it was the only space with roof access, which created problems when messy new soil or compost was needed. In 2009 the roof was completely replaced and a dedicated entrance was created. The roof also provided other insights: exposure to sun and wind create extreme growing conditions that dry out beds quickly—regular mulching and deep watering help; microclimates can be created by using trellises or other garden structures; gardening can start a few weeks earlier in spring than it could at ground level, extending the growing season; and, most important, not all plants grow well on the roof—heat-loving plants such as tomatoes, peppers, and basil do very well, while spinach, peas, and beans are less successful. Other educational projects include the New York City public schools (see page 174) and the Gary Comer Community Center in Chicago (see page 180), an oasis for inner-city youth.

Green Roof Technologies
Increased interest in green roofs have led to advances in technology. An entire industry has sprung up that specializes in lightweight growing media, filter cloths, roofing membranes, plant containers, and plant stock. However not all of this technology is suited to food production. Much of the research has focused on developing lighter, thinner green roof systems at a reduced cost that have minimal impact on the building structure, or on addressing concerns about leakage and resultant liability. Food production generally requires thick growing media and a more integrated approach. Green roofs consist of several layers: vegetation, a growing medium, a membrane, a drainage layer, a waterproofing membrane, and a support system. They are often categorized by their system type—extensive, simple-intensive, intensive, container, or greenhouse.

Extensive green roofs have shallow soil depth, typically 1 to 6 inches, and are primarily planted with succulents, mosses, grass, or herbaceous plants that are drought-tolerant or low maintenance. They add relatively little weight to a roof—typically 20-30 pounds per square foot when wet. Such limited soil thickness and water storage capacity reduces the extensive green roof's ability to grow food, with one notable exception being True Nature Foods in Chicago as discussed above.

Simple-intensive green roofs have a thicker soil layer and usually include grass, herbaceous plants, wild shrubs, and at times some vegetables and herbs. The soil depth is between 6 and 20 inches. Projects such as the RISC Forest Garden and the Gary Comer Youth Center are good illustrations of roofs of this type. This system provides storm water retention and wildlife habitat, which, when coupled with vegetable production, has the potential to become a small ecosystem of its own.

Intensive green roofs are characterized by thicker soil layers and, often, a mixture of hard and soft landscaping or planters. These require regular maintenance, have significant impact on the building's support structure, and are the most expensive but also the most suited to productive gardening.

Container roof systems have emerged as a popular, alternative approach to full horticulture systems. Containers can often be used on existing roofs, and they offer flexible design options. Simple, readily available containers such as car tires, curbside recycling bins, and buckets can be used to create inexpensive, informal gardens on virtually any roof or balcony. Custom-designed containers, such as those by Alternatives and Biotop, integrate developments in irrigation systems and soil replacements to reduce structural loads. Many container systems are modularized to allow the growth of a wider range of produce—these can be joined together to create a network of self-watering gardening containers as gravity pulls liquid down from topmost modules to water those below. These are often successful on existing roofs if they are placed above structural columns. The planters used at Uncommon Ground (see page 168) and the Fairmont Hotels (see page 188) illustrate container systems. Greenhouse spaces are very adaptable, but can be expensive to construct.

Conclusion

In dense urban locations, space is increasingly at a premium and urban agriculture has to compete with many other land uses. Consequently the potential of walls and roofs for generating resources—energy, water, or food—continues to be explored. While vertical components such as walls have potential for growing food, productive green walls are not yet common in modern urban buildings. Green roofs, however, are becoming established for a variety of reasons, and with these come opportunities for their use as productive spaces. Although the various recent municipal incentive programs to encourage the construction of green roofs are usually driven by the need to reduce water disposal infrastructure and the urban heat island effect, they can also spur the production of food in the heart of cities. As more municipalities consider food issues and develop strategies for addressing local food security, they are beginning to recognize the connection with green roof bylaws and incentives, both at the scale of individual or community plots on the roofs of homes or condominiums and of large commercial farms producing food for sale.

Productive roofs are now seen as part of an integrated approach to urban food production. Visionary projects such as Ravine City in Toronto (see page 30) and realized projects such as the Edible Campus in Montreal (see page 94) all include productive roofs as part of a comprehensive strategy for more sustainable urban living. Although the additional costs of green roofs can be significant, food production provides an opportunity to generate income from rooftop growing technologies, and it is likely that a variety of products and approaches will be developed in the coming years to facilitate wider adoption of food-producing areas on roofs.

NEW YORK CITY ROOFTOP FARMS:
EAGLE STREET AND BROOKLYN GRANGE

BEN FLANNER, ANNIE NOVAK, AND GOODE GREEN | GWEN SCHANTZ AND BROMLEY CALDARI
BROOKLYN, NEW YORK

New York City's outer boroughs have emerged as an epicenter for urban farming activity. Of particular note are two commercial, open-air, rooftop farms, Eagle Street Rooftop Farm and Brooklyn Grange, which were established in 2009 and 2010, respectively. The individuals who initiated both projects included professionals who wanted to change their career paths to become farmers. In both cases, the founders sought to combine the benefits of a green roof with urban agriculture—but their intent differed in terms of scale, orientation, and their relation to the community.

Eagle Street Rooftop Farm

Eagle Street Rooftop Farm is a 6,000-square-foot (560 square meter) organic vegetable farm located on the roof of a three-story industrial warehouse in Greenpoint, Brooklyn, overlooking the East River. It is the result of brainstorming that started in fall 2008 between several partners: Ben Flanner, a former engineer and e-trader turned farmer; Annie Novak, a farmer with six years of prior agricultural experience;[11] the owners of Broadway Stages, a Brooklyn-based sound/stage company with a history of community investment; and the green roof construction firm Goode Green, whose owner Chris Goode was already exploring how to use its green roof knowledge to start a farm. Broadway Stages owns the warehouse and financed the installation of the green roof in 2009 at a cost of approximately ten dollars per square foot, which included the development of the budget and crop plan as well as the cost of construction and equipment.

After obtaining the approval of a building engineer, 200,000 pounds of Rooflite Intensive, a lightweight, water-retaining soil mixture of compost, rock particulates, and shale, were hoisted to the rooftop with cranes. Volunteers then laid out the growing medium on a base system of polyethylene, drainage mats, and retention and separation fabrics. There are sixteen north-south facing beds, up to four feet in width and six inches thick, divided down the middle by a single long

ABOVE AND RIGHT: Views of Eagle Street farm with the New York City skyline in the background showing planting, left, harvesting, center, and watering, right.

aisle filled with mulched bark. In its first year, the Eagle Street Rooftop Farm used an irrigation system of black plastic drip lines. Since 2010 it has used city tap water for seedlings only, relying otherwise on rain, a rainwater catchment system, and drought-tolerant crops. Honey is harvested from beehives on an adjacent roof.[12]

The farm is now run by Annie Novak as the head farmer with a team of trained interns, urban farming apprentices, and staff from Growing Chefs, a group of agriculturalists and educators. Each week during the growing season, the farm relies on volunteers to assist with seasonal tasks, harvesting, and composting. Eagle Street Rooftop Farm has also operated a small community-supported agriculture program since 2010. In addition, the harvest is sold at a seasonal onsite farm market on Sundays, and to neighborhood restaurants. In partnership with Growing Chefs, the rooftop farm hosts a range of educational and volunteer programs including seed-saving and composting workshops.

ABOVE: Laying out irrigation systems on the roof at Brooklyn Grange.
ABOVE RIGHT: Preparing the roof at Brooklyn Grange to receive its topcoat of soil.
RIGHT, BELOW: An overview of the Brooklyn Grange rooftop.

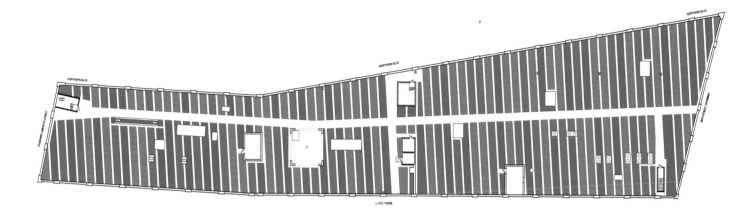

Brooklyn Grange

In late 2009, Ben Flanner moved on from Eagle Street to set up a much larger rooftop farm covering nearly an acre. Together with the team that created the backyard farm for Roberta's Pizza in Brooklyn,[13] he began this commercial, organic farming business. Brooklyn Grange was eventually established on a site in Long Island City, Queens, and covers over 40,000 square feet (3,700 square meters) on the roof of a six-story building.

It took only three weeks in the spring of 2010 to transform the vacant rooftop into a functioning commercial venture. Plans were filed with the Buildings Department to change the official use of the rooftop from "unoccupied" to "agricultural." Permit requirements included a structural analysis showing that the weight of the farm would be within the limits of the structure. The analysis revealed that the existing roof, which was a reinforced concrete slab, did not need additional support. Parapet walls and fire escapes were also examined and approved.

A crane was used to hoist truckloads of Rooflite Intensive soil mix to the roof. The 360 sacks of soil weighed over one million pounds total. The farm was installed at a cost of five dollars per square foot. Volunteers raked the mix over drainage and protective material, creating an 8-inch layer of soil. Volunteers then began planting thousands of seedlings to cultivate a wide variety of vegetables, including salad greens,

tomatoes, peppers, eggplants, cucumbers, squash, and kale, each selected for its ability to thrive in the sunny, windy conditions of an open city roof. The produce is sold at two markets at the building on Tuesdays and Thursdays and at Roberta's Pizza in Brooklyn on Sunday afternoons. A significant amount of produce is also sold to restaurants around New York City. Plans are being developed for expanding its direct markets in its second year of operations.

Both rooftop farms are privately owned and operated enterprises that are publicly accessible. While Eagle Street Farms is more community-oriented, with regular workshops for school groups and residents, Brooklyn Grange is a for-profit business that aims to demonstrate commercial opportunities for urban agriculture. Its rooftop site is much larger than Eagle Street's; the scale selected was based on a calculation of the size of farm that could both pay rent and provide a salary for at least one farmer. The careful start-up process of both farms in finding the appropriate team for the development of a commercial enterprise reveals the importance of networking and identifying realistic requirements for economically sustainable rooftop food production. Many aspects of both designs were carefully evaluated, from ensuring the roof could support the weight, to enabling access and solving safety issues, to planning for heavier elements to be placed over load-bearing structural beams. Design thinking pervaded the creation of these farms.

ABOVE: Brooklyn Grange's basic planting plan showing how much of the roof surface is devoted to production.
BELOW: The roof supports traditional building needs, such as the HVAC system, along with plants.

NEW YORK CITY ROOFTOP GREENHOUSES:
FOREST HOUSE AND ELI ZABAR'S VINEGAR FACTORY
BRIGHTFARM SYSTEMS AND BLUE SEA DEVELOPMENT | ELI ZABAR
NEW YORK, NEW YORK

Eli Zabar's Vinegar Factory and Forest House share some common characteristics: both New York City–based projects are creations of for-profit corporations, feature greenhouses on their rooftops, and are constructed with the intention to maximize local food production for sale. While the former is a cluster of classic greenhouse structures erected in 1995 on top of the supermarket that it supplies directly, however, the latter is a state-of-the-art hydroponics-based greenhouse to be built above a newly constructed, multifamily, affordable housing residential building.

Eli Zabar's Vinegar Factory

Eli Zabar bought a former vinegar factory on Manhattan's Upper East Side in 1993. He transformed the 9,000-square-foot (840-square-meter), two-story brick building into a gourmet marketplace that includes a bakery, a butcher, and other stores offering products made on site. Two years later, he decided to add greenhouses on the rooftop to grow fruits and vegetables for sale in the market below. His motivation was simply to reduce expenses and to be able to sell

fresh, local produce for a longer season while making use of the waste heat from the bakery below. Now greens, tomatoes, berries, and figs are grown and offered year-round.

A steel superstructure was installed atop the Vinegar Factory roof, which covers approximately half an acre, along with four greenhouses. The largest measures 40 by 100 feet (12 by 30 meters). Two full-time workers—with some help from other staff members—cultivate produce traditionally, that is, in soil. This choice

is not only because Eli Zabar believes that products grown in soil taste better than those grown hydroponically, but also because organic discards from the market like bread, fruit, and vegetables can be turned into compost used in the greenhouses, thus disposing of waste while producing enriched soil at no extra cost to the operation. Zabar is nevertheless considering the use of hydroponics in the future.

Forest House

Forest House, an affordable housing complex planned by Blue Sea Development Corporation for the South Bronx, will feature a 10,000-square-foot (930-square-meter) commercially run, hydroponic greenhouse on its rooftop, designed by BrightFarm Systems. The planning process began in 2008, with construction scheduled for the end of

ABOVE AND OPPOSITE: Renderings of the
proposed greenhouse for Forest House include spaces
for production, packing, and administration.
RIGHT: The newly constructed residential building
viewed from street level.

2010. The cooperation between these partners is intended to integrate emerging environmental technologies into affordable housing developments in New York City, while making some of the food produced available at low cost to neighborhood residents through a local nonprofit food cooperative, in an attempt to counter the presence of food deserts in the South Bronx.

The proposed greenhouse will make use of excess heat generated for the building's residential units. It will also capture the water runoff, estimated at almost 200,000 gallons (750,000 liters) per year, from the vast greenhouse roof. The greenhouse will be operated year-round, with an annual production estimated at 100,000 pounds of vegetables—half tomatoes and half lettuce—or roughly $400,000 worth of produce. The greenhouse will not be accessible to residents, but will be run by a commercial company.

Both of these projects concentrate not only on maximizing food production but also on simply reducing business expenses by capitalizing on readily available, sustainabile resources, like excess heat, to warm greenhouses for production in the colder months. These two commercially run greenhouses—despite their many differences—represent ways in which urban farming can be integrated into existing business models, rather than conceived of as start-up ventures like many other businesses in the urban agriculture sector.

BELOW: A steel structure supports the greenhouses and integrates the piping system.
BOTTOM: Herbs are cultivated in long beds.
OPPOSITE, ABOVE: Two greenhouses sit above retail spaces and blend seamlessly into the urban fabric of the neighborhood.
OPPOSITE, BELOW: A view into one rooftop greenhouse reveals row upon row of tomato plants.

UNCOMMON GROUND RESTAURANT

MICHAEL CAMERON
CHICAGO, ILLINOIS

Thanks to Chicago's Green Roof Grant Program, the city is home to over 200 green roofs, covering 2.5 million square feet. Green roofs appear at City Hall, the Apple store, and a McDonald's, for example, but not all of the city's green roofs produce food. However, some interesting productive green roofs are sprouting up around the city, both for community and commercial projects. The productive roof at Uncommon Ground restaurant's roof is an interesting example of a small-scale commercial operation and was designated the first certified organic rooftop farm in the United States in 2008.

Uncommon Ground is a community-based restaurant in Chicago that has pledged to use local, sustainable, and organically produced food. For many years the restaurant has supported a local-supplier approach, establishing relationships with farmers from the Great Lakes region who follow sustainable and organic methods. In 2007 when expanding to a second location, they took the local supply approach to a new level, constructing a productive green roof as part of a major renovation of the new site's 100-year-old building to provide organic produce for use in the restaurant below, and to educate the community about how to grow organic food on urban roofs. The roof has a total area of 2,500 square feet (230 square meters), of which

650 square feet (60 square meters) are covered with planters.

Construction of the rooftop garden required significant reinforcement of the existing building. The foundations were strengthened and the old timber roof supports were replaced with new steel columns and beams to bolster the brick load-bearing walls and carry the additional load of six tons of soil that volunteers carried up to the roof. Sustainable materials were chosen where possible, including recycled plastic and wood for the roof decking. Other features of the roof include five 4-by-10-foot solar thermal panels that can heat up to 70 percent of the restaurant's water, a seating area, workstation, and store. Materials and components were

reclaimed where possible and low emission finishes were specified.

The plants grow in twenty-eight planter boxes built from cedar and steel for durability and ease of use. They can be covered by a cold frame to expand the growing season and use a digitally programmed irrigation system for water efficiency. Excess water from the roof is collected in rain barrels and used to water the ground-level garden. The perimeter planters were designed to be 42 inches high—sufficient to meet city code requirements as railings. In the central area, ten movable, 10-by-4-foot planters can be rearranged to suit varying conditions and requirements. All these include trellises to support plants such as tomatoes, cucumbers, pole beans, and peas. Twelve subirrigated EarthBoxes are also used, making farming more efficient by allowing nutrients to move from more highly concentrated areas to less-concentrated areas. The plastic covers

TOP, ABOVE: Solar panels heat 70 percent of the restaurant's water.
ABOVE: A variety of planters are used.
RIGHT: Stairs at the building's rear give access to the rooftop garden.
OPPOSITE: A view over the garden and into the surrounding historic neighborhood.

on the EarthBoxes reduce the water evaporation rate from the soil to keep it saturated. The restaurant serves fresh vegetables from the rooftop garden, including sweet and hot peppers, eggplant, lettuce, heirloom tomatoes, collard, chard, kale, radishes, beets, okra, spinach, fennel, mustard, bush beans, and shallots. The roof also includes four beehives that produced forty pounds of honey in the first year, and a large variety of herbs such as rosemary, thyme, chives, garlic chives, tarragon, sage, parsley, dill, mint, lavender, basil, anise and hyssop.

The green roof at Uncommon Ground is part of a holistic strategy to minimize the impact of the restaurant's activities and serve nutritious, tasty food. Even the grease from the kitchen is converted to biodiesel. This strategy has helped to transform this small restaurant into a hub of community activity with a weekly farmers market, educational events, and music programs. It demonstrates the economic viability of a sustainable approach to serving food.

CARROT GREEN ROOF
TAFLER RYLETT ARCHITECTS, NATVIK ECOLOGICAL, AND CARROT COMMON
TORONTO, ONTARIO, CANADA

The main tenant of Carrot Common, a building in Toronto's Greektown neighborhood, is a worker-owned cooperative grocery store called The Big Carrot. The complex also includes retail and dining establishments, office space, and meeting rooms. With a leafy courtyard facing the small-scale yet busy commercial street of Danforth Avenue, the spot is a welcoming space to sit, meet friends, or shop. The owners took the opportunity to create a flagship living-roof garden and urban agriculture venture atop this unique space in the heart of Toronto.

The Carrot Common rooftop was used as a community space for many years. From 1989 to 1996, it functioned as a restaurant patio. A design charrette for a new roof garden[14] resulted in significant transformations. The roof became a green oasis, with deck areas and diverse plantings—including edible plants— growing in containers. In 2008 Carrot Common undertook a major reroofing project. At that time, many groups expressed interest in reinstituting the roof garden with a focus on sustainable practices. A second design charrette, this time with landscape architecture students from the University of Guelph, and Natvik Ecological, a consulting firm specializing in green roofs, both provided preliminary concepts for a productive green roof and community space. Accordingly, the existing roof structure was resurfaced with a membrane chosen for its compatibility with green roof systems.

It took two years of planning, designing, and consulting with many stakeholders and community groups[15] to arrive at the final design for the Carrot Green Roof & Garden Project. The main roof covers approximately 10,000 square feet; about 8,000 square feet accommodate a variety of community-related uses and structures. An extensive green roof using native plants, as well as intensive vegetable and herb gardens now cover the roof. A large deck was added to provide gathering space. The new roof also includes a rainwater harvesting system, a composting system tied to the Big Carrot food store below, a solar thermal power system, container gardens, living walls, screens, and other growing systems for vertical food production. At the heart of the roof is a covered but open-air community kitchen for community meals and cooking demonstrations.

Carrot Cache, the Big Carrot's non-profit arm, and its partners have made a goal of delivering environmentally related educational programming and exhibits to the immediate neighborhood as well as the wider community, partly by hosting workshops on techniques for waste management, rainwater harvesting, soil substrate testing, and vegetable growing methods. In this way, the Carrot Green Roof garden will become a learning hub for sharing knowledge and experience

TOP RIGHT: Food production as it existed on the roof before the 2008 design charrette.
BOTTOM RIGHT: Community members drawing a mandala on the roof's vast open space, before 2008.
OPPOSITE: A sketch for the stairway that invites the public up to the programming on the new roof.

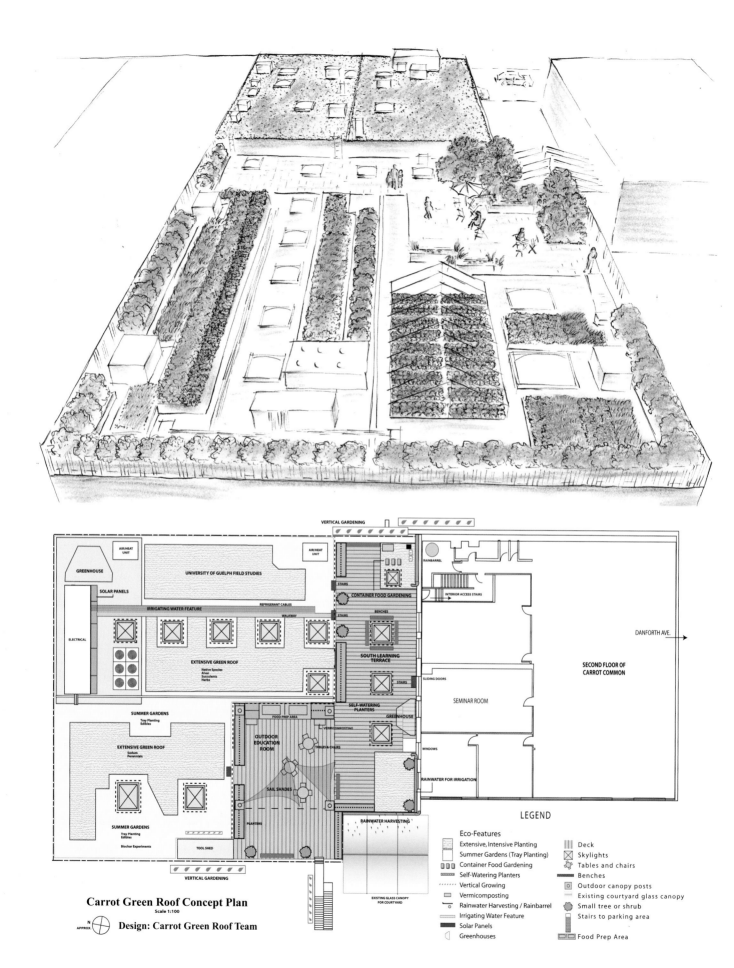

Carrot Green Roof Concept Plan

Scale 1:100

N APPROX.

Design: Carrot Green Roof Team

VERTICAL GARDENING

AIR/HEAT UNIT

GREENHOUSE

SOLAR PANELS

IRRIGATING WATER FEATURE

REFRIGERANT CABLES

WALKWAY

ELECTRICAL

UNIVERSITY OF GUELPH FIELD STUDIES

AIR/HEAT UNIT

RAINBARREL

INTERIOR ACCESS STAIRS

STAIRS

CONTAINER FOOD GARDENING

BENCHES

SOUTH LEARNING TERRACE

STAIRS

DANFORTH AVE.

SECOND FLOOR OF CARROT COMMON

EXTENSIVE GREEN ROOF
Native Species
Alvar
Succulents
Herbs

SELF-WATERING PLANTERS

GREENHOUSE

SLIDING DOORS

SEMINAR ROOM

SUMMER GARDENS
Tray Planting
Edibles

FOOD PREP AREA

VERMICOMPOSTING

EXTENSIVE GREEN ROOF
Sedum
Perennials

OUTDOOR EDUCATION ROOM

TABLES & CHAIRS

WINDOWS

SAIL SHADES

RAINWATER FOR IRRIGATION

SUMMER GARDENS
Tray Planting
Edibles

PLANTERS

RAINWATER HARVESTING

Biochar Experiments

TOOL SHED

VERTICAL GARDENING

EXISTING GLASS CANOPY FOR COURTYARD

LEGEND

Eco-Features

- Extensive, Intensive Planting
- Summer Gardens (Tray Planting)
- Container Food Gardening
- Self-Watering Planters
- Vertical Growing
- Vermicomposting
- Rainwater Harvesting / Rainbarrel
- Irrigating Water Feature
- Solar Panels
- Greenhouses

- Deck
- Skylights
- Tables and chairs
- Benches
- Outdoor canopy posts
- Existing courtyard glass canopy
- Small tree or shrub
- Stairs to parking area
- Food Prep Area

about green roofs and urban gardening. Space for experimentation was purposely included on the rooftop; this includes an area that will serve as a test zone for green roof technologies designed for northern climates.[16] Carrot Cache commissioned specific studies in preparing the new roof, including investigations into how the green roof could reduce peak storm water runoff, the feasibility of rainwater harvesting, and a waste audit of the Big Carrot to explore "vermi-stabilization" as a way to significantly reduce waste processing costs and generate a revenue stream from worm castings for use as an organic fertilizer and soil enhancer.[17] Experimental use of a biodome as a greenhouse is also planned.

Such explorations of integrated technologies in combination with educational and demonstration activities to promote diverse approaches to a green lifestyle are encouraged by Carrot Common tenants and The Big Carrot owners and staff. As a model of such practices, the Carrot Green Roof Project will be a resource for ideas on multifaceted aspects of sustainable living, combining the extensive green roof and productive roof approaches that have until now been largely seen as divergent approaches to "green" rooftop use. The project is timely, materializing just as the City of Toronto's recent Green Roof Bylaw requiring green roofs on all buildings with floor areas of 2,000 square meters or larger goes into effect. This demonstration and teaching community space shows that roofs can be amenities that fulfill many agendas and that they can serve more than the traditional purpose of providing shelter from the elements. In this way, this project shows how urban spaces like rooftops can seamlessly integrate production, education, and community activities.

BELOW, LEFT AND RIGHT: Construction photos showing installation of decking and newly planted green roof, autumn 2010.
OPPOSITE, TOP: Nativk Ecological's 2008 proposal for the redesigned roof.
OPPOSITE, BELOW: Tafler Rylett's concept plan for the fully redesigned roof showing the high percentage of total area dedicated to sustainable initiatives.

NEW YORK ROOFTOP SCHOOL GARDENS: P.S. 41, 64, AND 333

JONATHAN ROSE COMPANIES, MARK VETTER, BARBARA NORMAN, VICKI SANDO,
HANDEL ARCHITECTS, STANTEC ARCHITECTS, BRIGHT FARM SYSTEMS, AND KISS+CATHCART
NEW YORK, NEW YORK

Widespread discussions of climate change and nutritional problems over the last few years have led many parents and teachers to push for the introduction of outdoor classrooms and gardens in schools. Given the scarcity of open land across New York City, attention has focused on rooftops as teaching spaces relating to food production and food systems in school facilities. As a result, the transformation of the rooftops of several Manhattan public schools is currently planned. Issues of building structure, access, funding, and maintenance have led the designs to differ in regards to the types of space, i.e., greenhouses or open-air gardens, and growing methods to be used.

Greenroof Environmental Literacy Laboratory

The planning process for the transformation of the upper roof of P.S. 41, in the West Village, into a 14,000-square-foot (1,300-square-meter) open-air rooftop garden was initiated in 2006. This project, called Greenroof Environmental Literacy Laboratory (GELL),[18] emerged out of a successful garden program that a parent, Vicki Sando, had started in 2003 in the elementary school's courtyard with support from the National Gardening Association's Adopt-a-School

Garden Program. She and another parent, Christopher Hayes, developed the concept for a green roof laboratory that provides students with space for education about plants and nature. The project seeks to give urban children "environmental literacy" by integrating the roof space into the curriculum through the study of ecology, urban agriculture, sustainability, chemistry, biology, mathematics, nutrition, and technology. It is hoped the roof will lower the school's maintenance costs by keeping the building cool in summer and warm in winter.

Because structural limitations allow for shallow soil depth only, the garden will be laid out with a system of light, firmly installed trays where sedum and herbs will be grown—the production of vegetables and fruits will remain restricted to the ground-level garden. The rooftop herb garden will provide students with experience in marketing by partnering with several restaurants. One-third of the roof is set aside as a nature sanctuary for birds and insects to demonstrate the importance of providing habitat for urban wildlife. The other portion includes separate growing plots for each grade. Students helped by teachers and parents will tend the rooftop garden during the school year while during the summer, workshops will be held for teacher training on how to use a garden to teach. The effort to raise $1.7 million dollars for the necessary engineering, architectural design, project management and construction has taken four years so far.[19]

TOP AND ABOVE: The current roof of P.S. 41, top, and a rendering of the planned garden.
OPPOSITE: Outdoor classroom space is integrated into the plan for the rooftop garden at P.S. 41.

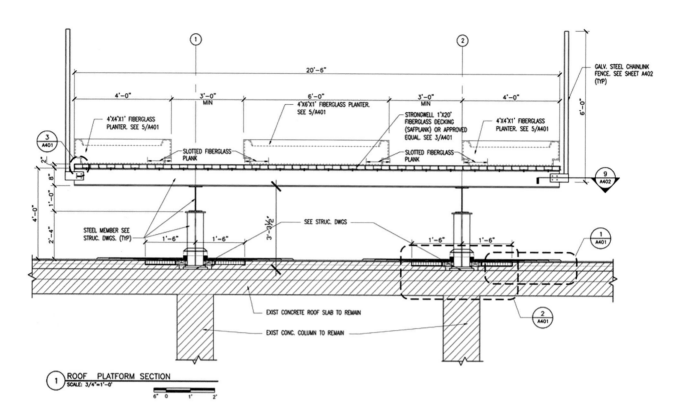

ROOF PLATFORM SECTION
SCALE: 3/4"=1'-0"

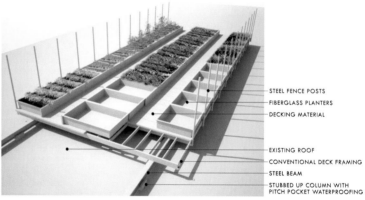

STEEL FENCE POSTS
FIBERGLASS PLANTERS
DECKING MATERIAL

EXISTING ROOF
CONVENTIONAL DECK FRAMING
STEEL BEAM
STUBBED UP COLUMN WITH
PITCH POCKET WATERPROOFING

TOP AND OPPOSITE, TOP LEFT:
Cross-section and rendering of the raised
roof structure.
ABOVE, LEFT: Rendering of the
roof garden with the city skyline in the
background.
ABOVE, RIGHT: Axonometric view
showing components of the roofing
system that distribute the load of the new
garden onto existing structural columns.
OPPOSITE, TOP RIGHT: Preliminary
plan for the redesigned P.S 64 roof.
OPPOSITE, BELOW: Raised beds at
child-friendly heights are planned for the
new roof garden.

Fifth Street Farm Project

The Robert Simon School complex, located in the East Village, houses three schools:
P.S. 64, Earth School, and Tompkins Square Middle School. A 3,000-square-foot
(280-square-meter) garden is planned on the roof of its three-story building. The
history and goals of the Fifth Street Farm Project are similar to those of GELL at P.S.
41, with a similar schedule. A grass roots organizational effort of parents, teachers,
and friends associated with the Earth School, which already has a small agricultural
program at ground level, is behind this proposal to enhance the learning resources and
environment of the school by creating a large green roof that is also a farmable rooftop
garden. The project will provide an average of 1,000 students per year with opportunities for critical scientific observation and hands-on work in ornamental gardening and
raised-bed vegetable cultivation. Their experiences in the garden will be integrated
into existing school courses on science and nutrition, and some of the produce will be
served in the cafeteria.[20]

The conceptual design for the rooftop garden was prepared by the school's staff
and parents in cooperation with the architect Michael Arad—whose child attends the
school—and then developed by Stantec Consulting into detailed construction documents. The first ideas for the rooftop consisted of a low-budget concept for placing

Legend
1. Planters with Fence
2. Extensive Green Roof
3. Roof Access Points
4. Meeting Space
5. Fenced Walkways
6. Fenced Entry/Exit Area
7. Roof Below
8. Phase Two Planters

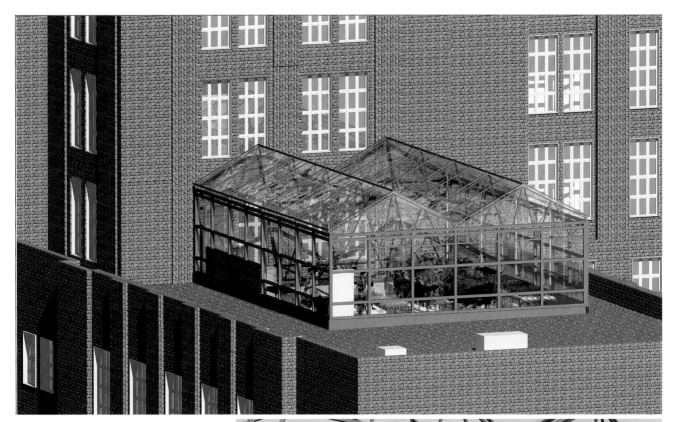

ABOVE AND RIGHT: The proposed Sun Works Center structure for P.S. 333.
OPPOSITE, TOP: The P.S. 333 rooftop greenhouse in the early phases of construction.
OPPOSITE, BELOW: The greenhouse nearing completion.

placing hundreds of plastic wading pools filled with soil on the roof, followed by a plan involving more costly prefab planters; these developed into a solution for a smaller deck system based on the way heavy equipment is often supported on roofs. The final plan involves cutting through the roof slab and stubbing up columns from a hallway in the center of the school, placing two long, steel beams on these extensions to serve as a foundation for a 20-foot-wide deck that rests about 4 feet above the actual 60-foot-wide roof slab. This system distributes the weight of the new rooftop deck across the existing building columns. The architects developed plans and details for minimizing the impact on the existing building while incorporating systems to address the safety and egress requirements associated with school use. In this way, they complied with New York City's School Construction Authority and the structural limitations of placement of an intensive green roof on an older building structure.

Sun Works Center for Environmental Studies

The Manhattan School for Children, P.S. 333, installed a 1,440-square-foot (130-square-meter) greenhouse laboratory on top of one of its roofs. Located in a three-story building on the Upper West Side, the project developed out of the school community's desire to create a more sustainable campus and environmentally focused curriculum.[21] The nonprofit organization NY Sun Works, a school-based successor to the Science Barge, is spearheading the project, which is hoped to serve as a prototype that can be replicated at other schools throughout New York City. The Sun Works Center for Environmental Studies will be a laboratory that provides about forty students with a year-round possibility for hands-on learning of science concepts, environmental sustainability, and food production. The greenhouse will also provide teacher education and professional development through collaborations with neighboring institutions, as well as after-school and weekend workshops for teachers, students, and members of the community.

BrightFarm Systems—NY Sun Works's for-profit counterpart—and the architecture firm Kiss + Cathcart were commissioned to design the center, which utilizes hydroponic greenhouse technology powered by renewable energy from solar panels and incorporates rainwater collection, evaporative cooling, aquaponics, vine crop systems, and integrative pest management. The lab also features raised-soil beds for sustainable food production with differentiated planting zones for individual and grade projects, a composting center, insect-hatching areas, a weather station, and an indoor classroom with a kitchen corner. The greenhouse structure and accompanying developments built on the school's rooftop required only minor alterations to the existing space to provide access to its terrace location. The greenhouse was erected and the growing systems installed in the fall of 2010.

New York City has seen a rapid rise of rooftop gardens in public schools; other promising examples include the Harbor School on Governors Island and the Food and Finance High School on the West Side of Manhattan. These were also initiated by parents and teachers who raised funds to improve the curriculum of their children's schools. With the support of certain elected officials, they have managed to come up with

substantial sums to install the gardens according to the School Construction Authority's requirements.

Mostly due to financial issues and existing building conditions, school gardens vary considerably in the approach taken in their design. In all cases, the construction requirements imposed by the authority and the difficult conditions associated with erecting complex growing systems on top of flat roofs of older buildings has led to design challenges that require sophisticated and expensive design responses. Despite these challenges, overall the New York City cases offer lessons for combining sustainable, year-round food production with education, while integrating various production approaches like hydroponic and aquaponic systems with other elements that enhance sustainability, such as solar panels and rainwater collection—merging nutritional literacy with environmental literacy.

GARY COMER YOUTH CENTER

HOERR SCHAUDT LANDSCAPE ARCHITECTS AND JOHN RONAN ARCHITECTS
CHICAGO, ILLINOIS

Food production and education go hand-in-hand particularly well. In locations where ground-level gardening may attract unwanted attention or be vandalized, rooftop gardens offer an interesting solution for secure growing space. The Gary Comer Youth Center is an example where the benefits of a protected roof space have been exploited to create a unique and beautiful project.

Located on an infill site close to train lines and surrounded by a main road and parking, the Gary Comer Youth Center occupies a disadvantaged neighborhood site in Chicago's South Side with little access to safe outdoor environments where children could learn about the natural world. Since 2006 the center has provided an outdoor classroom space and a safe haven for children and seniors from the community to learn about plants and food. Hoerr Schaudt Landscape Architects designed the 8,160-square-foot (760-square-meter) green roof, working closely with the building's architect, John Ronan. It is located on

top of a gymnasium and café, and is surrounded on all sides by circulation corridors and classrooms, which provide shelter and some seclusion from the outside world. As students move through the building's third floor they can see the garden beds through floor-to-ceiling windows labeled with bold graphics that identify various plants. Large round skylights provide daylight to floors below and act as large sculptural elements in the garden that contrast with the straight lines of planters. Recycled materials such as plastic lumber created from recycled milk containers and recycled tire pavers are used as structural elements.

The garden is designed as a series of long strips of plant beds that hold between 18 and 24 inches of soil, allowing organic cabbage, sunflowers, carrots, lettuce, and strawberries to be grown. In one year the garden has grown over 1,000 pounds of food, consumed at the center's café and local restaurants. Growing conditions on the roof are markedly different from what they would be at ground level. Solar gains and heat loss from the gymnasium below mean that it is possible to keep some planters above freezing even during the Midwestern winter and extend the growing season; herbs and even spinach can be grown under tentlike structures year round.

In an area of little biodiversity, the garden has become an oasis not only for children and seniors but also birds, bees, worms, caterpillars, and other wildlife that was seldom seen in the area previously. A full-time garden manager employed by the center has developed creative ways to use the space for horticultural learning, environmental awareness, and food production. Each group of students gets a chance to experience and learn about the whole seed-to-table cycle and enjoy the produce they have grown. This green roof is a model for exploiting traditionally underutilized space to the benefit of the community. It is a unique space of respite from the environment below, and introduces children to the wonder of growing plants for food while teaching them about nutrition and the value of fresh food.

BELOW: Round skylights punctuate rows of neat vegetable and flower beds on the roof, allowing sun to filter to the gymnasium floor below.
OPPOSITE: Corridors and classrooms ring the garden, fostering a sense of temporary seclusion from the outside world.

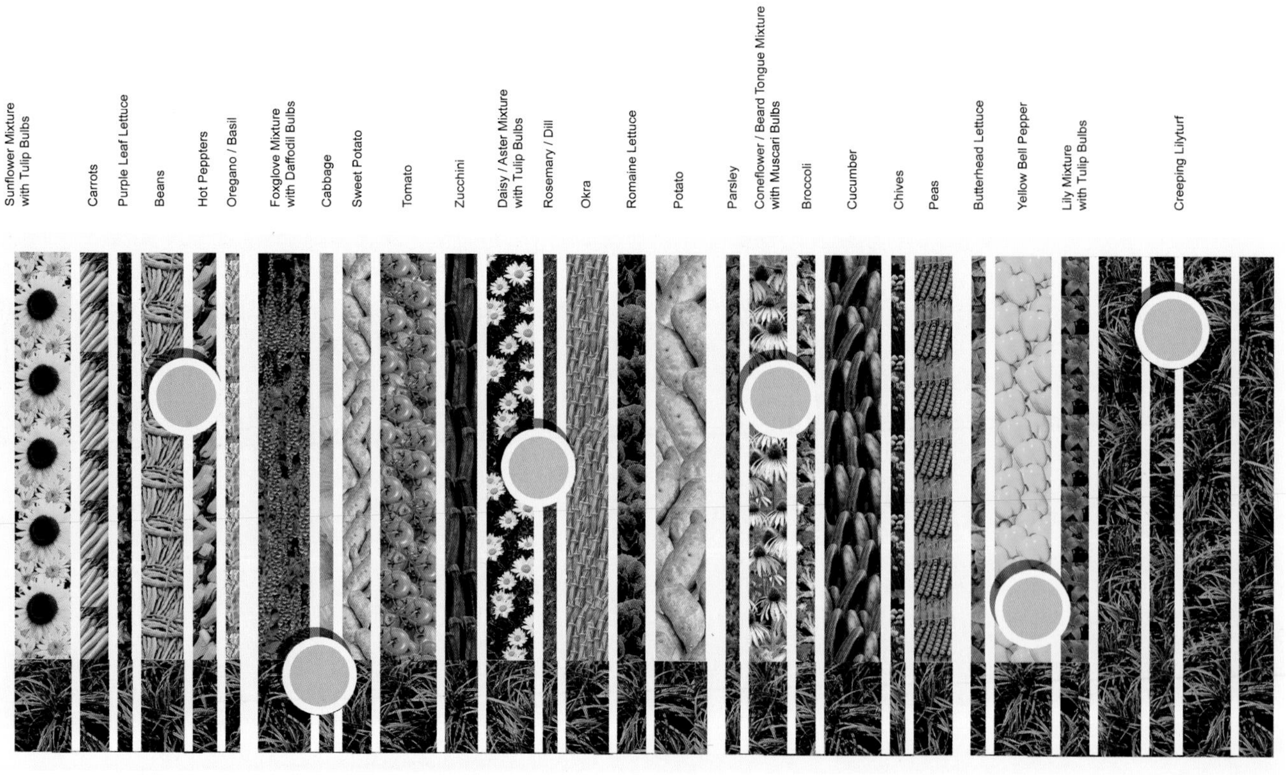

Sunflower Mixture with Tulip Bulbs

Carrots

Purple Leaf Lettuce

Beans

Hot Peppers

Oregano / Basil

Foxglove Mixture with Daffodil Bulbs

Cabbage

Sweet Potato

Tomato

Zucchini

Daisy / Aster Mixture with Tulip Bulbs

Rosemary / Dill

Okra

Romaine Lettuce

Potato

Parsley

Coneflower / Beard Tongue Mixture with Muscari Bulbs

Broccoli

Cucumber

Chives

Peas

Butterhead Lettuce

Yellow Bell Pepper

Lily Mixture with Tulip Bulbs

Creeping Lilyturf

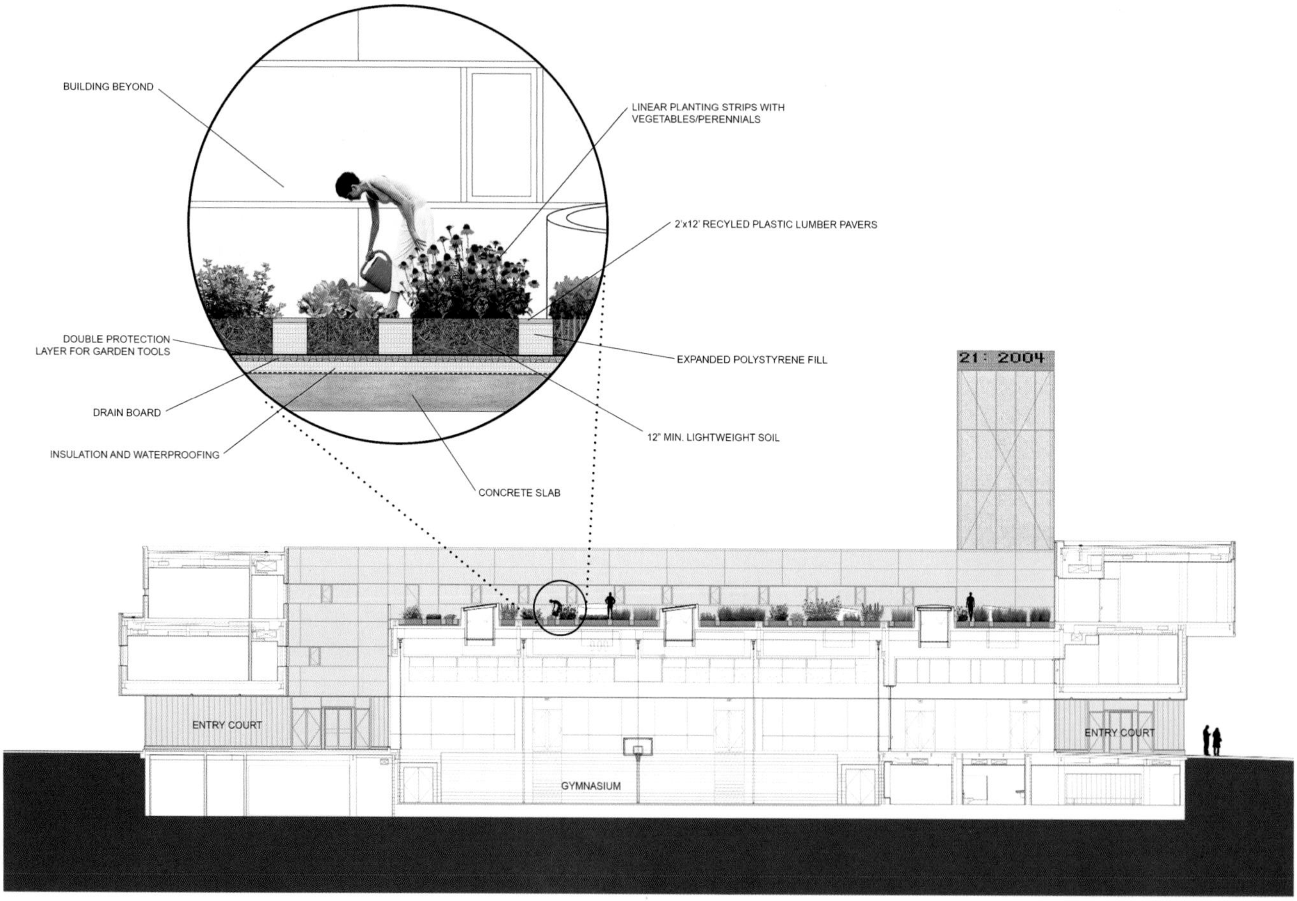

BUILDING BEYOND

LINEAR PLANTING STRIPS WITH VEGETABLES/PERENNIALS

2'x12' RECYLED PLASTIC LUMBER PAVERS

DOUBLE PROTECTION LAYER FOR GARDEN TOOLS

EXPANDED POLYSTYRENE FILL

DRAIN BOARD

12" MIN. LIGHTWEIGHT SOIL

INSULATION AND WATERPROOFING

CONCRETE SLAB

21: 2004

ENTRY COURT

ENTRY COURT

GYMNASIUM

ABOVE: The garden, viewed from above, appears as an oasis in a part of Chicago that has long been identified as a "food desert."
OPPOSITE, ABOVE: Vegetables alternate with rows of ornamental flowers to create a tapestry that is colorful and varied.
OPPOSITE, BELOW: A north-south cross-section of the building and garden beds.

BRONXSCAPE, LOUIS NINE HOUSE

NEIGHBORHOOD COALITION FOR SHELTER AND PARSONS THE NEW SCHOOL FOR DESIGN
BRONX, NEW YORK

The Neighborhood Coalition for Shelter (NCS), a nonprofit organization providing homeless men and women with housing and support, commissioned Parsons The New School for Design in early 2008 to create a community space on the roof of a new four-story apartment building—now known as Louis Nine House—to house forty-six young adults reaching age eighteen, and therefore aging out of the foster care system. The building,[22] which opened in early 2009, seeks to combine housing with educational programs that teach residents the necessary life skills for independence. The NCS program for the 4,500-square-foot rooftop provides options for the residents to learn and participate in the food cycle from planting crops to cooking, consuming, and finally composting.

David J. Lewis leads Parsons's design-build studio, known as The Design Workshop, which provides pro bono design and construction services to nonprofit organizations.[23] A group of fifteen students taking part in the 2008 workshop developed what they called the bronXscape proposal: a rooftop garden and kitchen pavilion that includes spaces for vocational training as well as leisure and recreation. The rooftop offers both communal gathering spaces and private zones.

The studio's collaborative design, developed in the first half of the spring semester, included an open-air roof garden with a covered, multifunctional pavilion in the center that provides a shaded area for working on seedlings at the start of the season and for food preparation at harvest time. Incorporated within this structure is a greenhouse serving as the anchor for the roof garden's productive spaces. During the second half of the spring semester, the students developed

construction drawings, cost estimates, schedules, and purchase orders to realize this project.[24]

While the building complex was still under construction, students designed the individual components as parts and built them off-site at Parsons, later moving the pieces to the Bronx and assembling them on the roof. To prefabricate the steel structure to accurately touch down onto existing load-bearing walls, the students calculated and coordinated all the structural loads and built a comprehensive 3-D computer model of the complex, down to every last nut and bolt.

This fabrication process was organized to allow the students to erect the structure without the use of special tools or heavy machinery, thereby also reducing costs. When the group entered the summer construction phase, the preparation and erection of the steel went quickly and efficiently—the majority of the steel structure was assembled in a week's time.

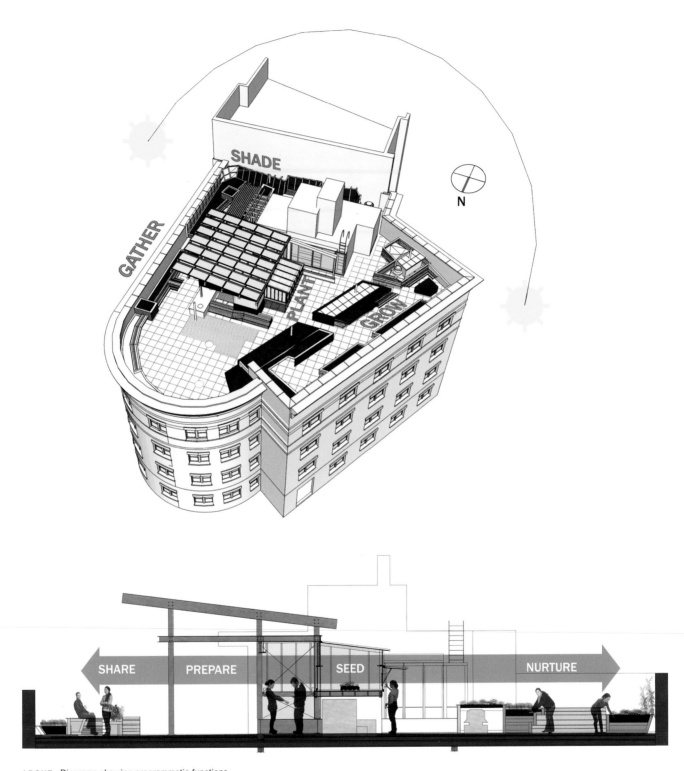

ABOVE: Diagrams showing programmatic functions intended for the completed rooftop garden.
OPPOSITE, LEFT: The completed, open-air pavilion provides space for cooking, eating, and entertainment.
OPPOSITE, RIGHT: A team of architecture students from Parsons The New School for Design, constructing the pavilion.

ABOVE: Staggered photovoltaic panels provide shade to the pavilion's main programmatic spaces while perforated screens wrap the pavilion's functional surfaces.

RIGHT: A construction sketch for the pavilion's frame.

OPPOSITE: The completed greenhouse features a series of panes that open wide to allow cross-ventilation in warmer months.

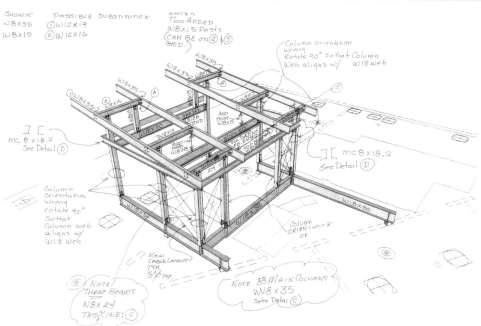

The ability of the roof to support the weight of soil was an important consideration. The proposed structural system for the new building was planned so that larger and deeper planting boxes were placed on top of the bearing walls only. This ensured that these areas would not exceed the allowable loads when the weight of planters, soil, moisture, and residents are combined. Engineered soil that weighs less than a normal soil mix was also used.

The rooftop design also had to accommodate existing plumbing vents and ventilation ducts. Many of the planters were placed around these mechanical components, and were slatted with black locust wood to allow adequate airflow to reach the ducts. The design also considered the effects of a taller building to the south and an existing elevator penthouse on available sunlight for crops.

The planted perimeter provides a lush atmosphere on the rooftop for the majority of the year. Integrated benches and lighting provide a safe, relaxed environment for socializing. The western side of the roof has planters for intensive gardening for use as vocational training requested by the client. Denser plantings integrated with bench seating on the eastern and northern sides provide space where residents can congregate and share communal meals. The southern part of the roof, which does not receive any direct sunlight, has a green wall with built-in benches and a relaxing shade garden to provide a cool spot during the hottest summer months. The vertical garden in this area was planted for texture, color, and dimension; its frames attach the wall to the roof while also masking vents and ducts.

The main pavilion occupies the center of the space, is open to the air but sheltered from rain, and provides surfaces for food preparation, eating, and entertainment. The interior is wired to accept plug-in appliances. Integrated illumination results in a well-lit space for use throughout the day and at night. Perforated panels provide enclosure, thereby improving protection from wind-driven moisture while allowing air to pass through the vents of the building's mechanical units.

A canopy and a trellis hang over the entire pavilion. A greenhouse is integrated into the pavilion, which sits above the existing air-handling unit and serves as the physical connection between the kitchen and the vegetable garden. It is intended for starting seeds in the winter. The operable greenhouse windows open during the summer months to provide cross-ventilation for the pavilion. A lockable tool shed provides storage for the gardening equipment that residents learn to use. A small sink is located near the preparation area for rinsing vegetables and cleaning up after meals.

The team introduced a photovoltaic shading device into the canopy that is directly integrated into the electrical grid. The steel structure supports an array of thirty-five photovoltaic panels tilted 10 degrees towards the south. At roughly 6 kilowatt hours peak output, the array enables the structures on the roof to offset some of the building's electric load.

For irrigation of the perimeter gardens, a water-catchment system was installed, which includes the surface of the canopy. Residents have access to natural compost bins that turn their organic waste into fertilizer for use in the perimeter gardens. Once seedlings started in the greenhouse mature, they are transplanted to these perimeter gardens, where residents water and nurture them. As the plants start producing vegetables, the residents can harvest the food and prepare meals as a community in the adjacent kitchen space.

The constructed garden fulfills the workshop's motto, "Share, Prepare, Seed, Nurture," as intended. Because the canopy is visible from street level and from the elevated subway running nearby, it makes a statement as a significant street presence for the client as well as the residents who use the roof. The design and construction of bronXscape succeeded in meeting the energy and sustainability goals of the overall complex, transforming the roof into an active urban oasis, a community space, and a retreat for the young residents who will soon transition into adult life.

FAIRMONT HOTEL GARDENS
FAIRMONT HOTELS AND RESORTS
VANCOUVER, BRITISH COLUMBIA AND TORONTO, ONTARIO, CANADA

Vegetables and a variety of fruits and herbs are thriving on the rooftops of many of the Fairmont Hotels and Resorts luxury chain, supplying their restaurants with the freshest possible ingredients. Chefs at many of their restaurants now cook with fresh organic herbs and exotic items like alpine strawberries and garnish dishes with edible flowers such as violet pansies picked from their own productive gardens. The Royal York in Toronto and the Waterfront Hotel in Vancouver, B.C., were the first two to participate in the urban farming project. Fairmont hotels in Boston, Montréal, Bermuda, and Washington are now following suit. These gardens are also pesticide- and herbicide-free. Six of the hotels have beekeeping facilities with impressive yields, too. These also improve pollination; the Fairmont bees seek pollen from flowers growing in their own rooftop gardens and feast on a variety of plants in ornamental gardens throughout the city as well.

Fairmont Waterfront Hotel, Vancouver, Canada

The twenty-story Fairmont Waterfront Hotel in Vancouver was constructed in 1991 with a green roof on the third-floor roof terrace, which can be seen from the guest rooms. The garden was conceived of as an amenity for guests from the start, and is a venue for parties and receptions. Originally planted with ivy, the southern part of the terrace was converted into an herb garden in 1994, consisting of eleven planting beds in various shapes.

In total, this project covers 2,100 square feet (195 square meters) of the roof. The soil consists of the roof's original soil mix supplemented regularly with organic soil. The garden harvest includes over sixty varieties of herbs, fruits, edible flowers, and vegetables. In 2008 bee-keeping was added, and now there are 390,000 honeybees making honey from local flowers. In 2009 they produced 485 pounds (220 kilograms) of honey for use in the hotel restaurants. Justifiably proud of its success, the beekeeper, Graeme Evans, conducts weekly garden and hive tours. People are not the only beneficiaries though—over ten species of local birds have been drawn to the garden, an example of reintroduced biodiversity.

Fairmont Royal York, Toronto, Canada

In 1998 a garden was created on the fourteenth-story rooftop of the Fairmont Royal York in Toronto with an initial investment of $4,000. This retrofit of a formerly unused roof above the build-ing's core created a 4,000-square-foot

TOP, ABOVE: Many of the guest rooms at the Fairmont in Vancouver look out over the semicircular garden at mezzanine level.
LEFT AND ABOVE: Paths meander through the vegetables, herbs, and flowers.
OPPOSITE: Plantings were designed with function as well as aesthetics in mind.

(370-square-meter) garden that now contains seventeen raised beds and twenty-three planter pots used to grow a variety of herbs and vegetables, such as lemon thyme, chives, rosemary, Italian red kidney beans, and cherry tomatoes. A horticultural expert at a local nursery, Mason House Gardens, provides advice and support and apprentice chefs tend the garden and harvest the crops on a daily basis, a job made easier by waist-high frames that hold the raised beds. Garden care has been fully integrated as a part of the kitchen's three-year apprenticeship program. In the summer of 2008, following the trend from Paris for apiarists to keep bees on the roof of the Eiffel Park Hotel and the Opera House, Executive Chef David Garcelon introduced beehives at the Royal York. The colony is now estimated at about 300,000 bees, tended by the Toronto Beekeepers Co-operative. Their honey is enjoyed by the guests and has also won prizes at Toronto's annual Royal Agricultural Winter Fair.

While the rooftop is not normally accessible to guests, the garden has evolved from a little-known project to support the kitchen and train its staff to a marketing asset for the hotel, which now gives weekly tours, in season, to restaurant patrons in combination with afternoon tea, and to interested groups on demand, adding a demonstration function to the garden. The space is fragrant, lush, and has a dramatic view of the city and Lake Ontario.

Other Fairmont Hotels

In 2010 the executive chef of San Francisco's Fairmont, jW Foster, introduced a new 1,000-square-foot (93-square-meter) culinary garden, as well as four beehives, to this downtown hotel. Lavender, rosemary, thyme, oregano, basil, chives, and

cilantro were planned crops. Before joining the San Francisco Fairmont, Foster had also established a 3,000-square-foot (279-square-meter) productive garden at the Fairmont Dallas, and had worked in partnership with Marshall's Farm, a honey-producing farm in Northern California, since 1993. In addition to helping the hotel restaurants offer a sustainable, local cuisine with pesticide-free ingredients, the gardens and their grand vistas can be enjoyed by guests.

The initiative to install productive gardens and keep bees is part of the Fairmont hotel chain's commitment to environmental protection and sustainability, and the movement is spreading. The Four Seasons hotels in Philadelphia and Chicago now grow herbs and vegetables on their roofs. New York's Gramercy Park Hotel also has a new initiative begun by sous-chef Dan Dilworth and restaurant manager Kevin Denton; they salvaged and reused a variety of containers to create an herb-and-vegetable garden irrigated with water captured in metal drums donated by an oil company.

Hotel restaurants are uniquely situated in buildings that their owners control, meaning roof access is guaranteed. These large-scale buildings have roof space adequate for gardens that can make a significant contribution to the food cooked in their restaurants. They also have the staff to be able to maintain large gardens. Hotel management must have a commitment to the project, but recent examples speak to its feasibility. The amazing settings often occupied by luxury hotels and the marketing draw for guests are clear arguments for designing hotels with accessible growing spaces for guests to enjoy.

GARDEN TABLE & CHAIR

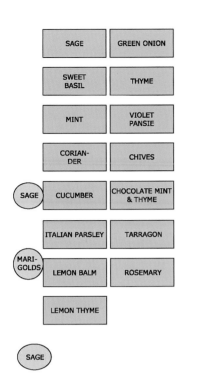

SAGE | GREEN ONION

SWEET BASIL | THYME

MINT | VIOLET PANSIE

CORIAN-DER | CHIVES

SAGE | CUCUMBER | CHOCOLATE MINT & THYME

ITALIAN PARSLEY | TARRAGON

MARI-GOLDS | LEMON BALM | ROSEMARY

LEMON THYME

SAGE

JACKMANII CLEMATIS/ MARIGOLD/SNOW PEAS

MANDE-VILLA | ENTRANCE

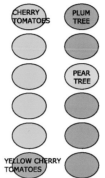

ITALIAN RED KIDNEY BEANS

CHERRY TOMATOES | PLUM TREE

PEAR TREE

YELLOW CHERRY TOMATOES

CHERRY TREE

SNOW PEA/ BALTIC IVY

MANDE-VILLA

LEFT: Planting plan for the Toronto Fairmont Royal York Hotel's roof garden.
BELOW: Waist-high raised beds were installed to make cultivation and care easier.
BOTTOM, LEFT: Beehives on the roof supply hotel guests with fresh honey.
BOTTOM, RIGHT: A break area for hotel employees overlooks the herb garden at the Fairmont Royal York.
OPPOSITE, LEFT: Executive chef David Garcelon in the Fairmont Royal York's garden.
OPPOSITE, RIGHT: The Fairmont Royal York's raised beds in full productive season.

BROWNSTONE ROOF GARDEN
JEFF HEEHS
BROOKLYN, NEW YORK

This four-story, 1910 row house in Brooklyn, New York, is typical of the residential brownstones in the Park Slope neighborhood. The Brooklyn Botanical Garden is a short walk away, a rich resource of beautifully designed parkland that also offers many workshops about horticulture, from choosing plants to composting. Park Slope is also the location of one of the best-known American food cooperatives, and has become a hub for food activism. Local resident Jeff Heehs is engaged in all these dimensions of the neighborhood, and it was in this context that he was inspired to create a rooftop vegetable garden in 2005. His design provides fresh food for himself and his friends, and serves as a green space with extraordinary views of the city. As the resident of the top-floor apartment in the building, he has exclusive access to the flat roof via a hatch.

The building's flat roof is 1,200 square feet (111 square meters) and accommodates some mechanical equipment and vents, with ample space for a garden. It was first designed with 180 square feet (almost 18 square meters) of planting area, plus walkways between beds. The vegetation grows in both shallow and deep commercial planter boxes that enable both extensive and irrigated intensive planting beds. These planter modules are also portable, meaning they can be easily arranged and give easy access to the cold-adhesive self-sealing rubber roofing membrane installed below. A wooden deck featuring a seating and outdoor cooking area was built above the membrane. Weston GreenGrid modular planters are placed between the walkways. While these are normally used as part of preplanted green roof designed for low-maintenance plantings such as sedum, Heehs adapted their 4-inch-deep, 2-by-4-foot units for ornamentals and herbs such as sedums, hens and chicks, allium, and lemon thyme. Eight-inch-deep modules hold weightier vegetables like salad greens, tomatoes, peppers, and green onions.

The modules include drain holes in raised channels, exit paths for water. The trays are lined with a filtering fabric to keep the irrigation system from clogging, and filled with a lightweight growing medium that holds in moisture and nutrients. The planted boxes are watered with a drip-tape irrigation system complete with a timer attached to a detachable garden hose that connects to an indoor water source in Heehs's kitchen, on the level below. This allows

him to leave the garden unattended on occasion. The growing medium includes compost and a mixture of Stalite—a commercial product—vermiculite, and perlite, lightweight aggregates that replace soil, greatly lowering the weight of the planter boxes and making it possible to grow vegetables on a roof not designed for heavy loads.

Heehs has continued experimenting with new plants and techniques. Each element has to be carefully considered, sourced, and pieced together to work as a system appropriate to the roof of a 100-year-old brownstone. The only access to the roof presently is by means of a ladder—during construction, the decking material was delivered with a hoist and Heehs brought soil and equipment up several flights of stairs and the ladder himself—and, second, roof loads from the added weight of a heavy snowfall atop this planter system could be a considerable burden on a structure not designed for such a combination. When filled with water, the growing mixture weighs about 28 pounds per cubic foot (136.7 kilograms per square meter), equal to a load of approximately 2 feet of wet snow on the same

surface; consequently, Heehs has to constantly clear the snow from the roof during storms. Clearly only the most dedicated gardeners will work under such circumstances.

Despite these difficulties, the green roof contributes in its small way to a healthier environment for Brooklyn, by reducing the heat-island effect for the city and food miles for the owner. Countless brownstones as well as other residential building types have at least some of the constraints identified here. It is important to develop technologies and strategies to make such gardens more achievable for the average resident. Building code implications are paramount; devising easily sourced, lightweight growing media and irrigating planters are only the first step, for as yet, parapets, water sources, and roof access are still challenges that many city dwellers cannot overcome.

ABOVE, LEFT AND RIGHT: Wooden paths allow the owner access to beds planted with a variety of herbs and vegetables.
OPPOSITE, LEFT AND RIGHT: The street facade of this quintessential Brooklyn brownstone does not hint at the bounty to be found on its roof, left.

COMPONENTS FOR GROWING

Large-scale conventional food production relies on vast expanses of rural land, the use of heavy machinery, and chemical fertilizers. These needs do not integrate well with a city's urban fabric. Cultivation and the rearing of livestock within urban areas—in and around buildings and on small, scattered sites—requires a very different set of processes, tools, components, and systems. Urban food production is usually small in scale, typically shares space with other land uses, and makes use of urban waste spaces such as vacant lots, roofs, walls, and balconies. It must incorporate readily available resources. Urban agriculture benefits from small-scale systems and components that enable opportunistic and intense cultivation in pockets of high-value spaces. These systems are strengthened by proximity to people or buildings, and can often be integrated with rainwater collection or a reuse of waste heat.

This section shows examples of ideas that can be applied in various situations and used individually or as part of a larger system or project to help produce food in urban locations. Some designs tackle the problem of prohibitively expensive soil remediation by creating container gardening systems, some address the problem of limited growing space by finding ways to use available space more intensively, and others suggest approaches appropriate to balconies and roofs that cannot support heavy planters. Green walls, various planter systems, livestock shelters, greenhouses, hydroponic systems, and mobile support structures are all examples of technologies that are effective in urban conditions. While some of these items are based on sophisticated science and state-of-the-art materials, many use off-the-shelf components or even salvaged components to provide creative urban agriculture solutions at low cost for a variety of challenges.

Most of these components can be placed individually, but sometimes a particular component can be adapted for use in a larger way that benefits a whole community. The soft planter bags and raised beds developed by What if: projects in London led to the Vacant Lot project, where many planters were artistically introduced to a single location—the success of this endeavor also led to their use in council housing estates elsewhere in the city (see page 220).

Salvaged Systems

Components used for urban cultivation vary in their complexity and/or use of technology. There are many examples of simple DIY technologies such as the reuse of plastic containers for growing plants, which requires only basic gardening skills and creativity. For instance, to increase food production in Colombo, Sri Lanka, residents, together with a team from McGill University's Minimum Cost Housing Group and ETC Energy, from Holland, developed a strategy of very small-scale growing spaces using recycled objects as planters.[1] Discarded tires, soda bottles, cooking oil containers, and soil-filled plastic bags are used there to grow vegetables and herbs. Recycled planters sit on railings, hang from eaves, and are placed between exterior walls and the street. Not only do these systems provide immediate access to fresh herbs and vegetables in terribly crowded conditions, but they fill a depressed urban space with soothing greenery. Similar strategies can easily be reproduced in similar developments worldwide as overcrowding increases.

In industrialized countries where soil contamination requires plants to be separated from the existing ground, gardeners have come up with creative approaches to low-cost, raised beds. Several were designed with accessibility for the elderly and the mobility-impaired in mind. Here, waste containers such as old packing crates, buckets, recycling boxes, and even bathtubs are used. Beds constructed of scrap lumber can also often be turned into cold frames to extend the growing season when fitted with old windows as "roofs" to create a warming, greenhouse effect on the interior. The greenhouse constructed by the owners of Roberta's

BELOW: Hollowed-out stalks of bamboo make inexpensive, flexible containers for plants in Colombo, Sri Lanka.

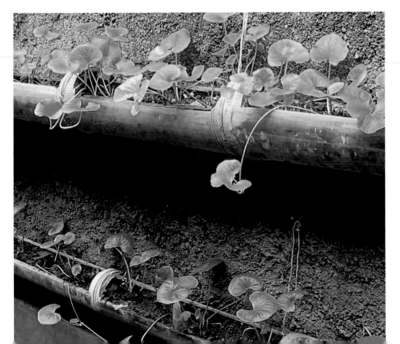

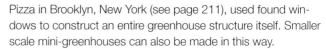

Pizza in Brooklyn, New York (see page 211), used win-dows to construct an entire greenhouse structure itself. Smaller scale mini-greenhouses can also be made in this way.

ABOVE AND RIGHT: The Molecular Kitchen, a series of portable carts based on a structure of recycled bicycle components, provides mobile, temporary food preparation and serving surfaces.
BELOW: A view of Growing Power's simple aquaponic system in Milwaukee, Wisconsin, devised from off-the-shelf components available at home improvement stores.

New Technologies

Sophisticated new materials and technologies have been adapted to many urban growing systems. New textiles create lightweight, hanging bags for cultivation; planters with soil-replacement growing media reduce crop weight, allowing them to sit on rooftops and in vertical configurations; and moisture-sensing irrigation systems minimize water usage while reducing labor needed for crop maintenance. Recently a variety of techni-cally sophisticated systems using rigid containers with water reservoirs in their base have been devised; these irrigate the soil in the container above through simple capillary action. Examples include the Alternatives plant container, the BioTop, and the Earth Box (see page 222). Some of these containers can even be combined or stacked into a series of modules appropriate for use on roofs or balconies, and many have aesthetically intriguing forms that make them attractive as decoration.

Hydroponic, aquaponic, and aeroponic technologies for growing plants without using soil as a nutrient source have revolutionized the way vegetation can be integrated into architecture, and have increased the potential productivity of building components such as walls, roofs, and double-skin glazed facades. NY Sun Works has introduced hydroponics creatively into several realized projects, including the educational Science Barge—a converted industrial ship in New York, (see page 86)—and greenhouse science labs on the rooftops of a number of New York City schools (see pages 174). Restaurants are increasingly growing herbs, fruits, and vegetables on site using hydroponic systems; the Bell Book & Candle in Manhattan already has implemented such a rooftop system successfully.

Some of the most interesting recent ideas combine new technologies with readily available or salvaged materials into new mechanisms that can be cheaply and easily implemented. For example, WindowFarms combines hydroponic technologies like pumps and nutrient-enriched water with salvaged empty water or soda bottles to create a system that can be mounted in virtually any window (see page 208). The Solar Bubble Greenhouse improves both traditional insulation levels and growing conditions by pumping liquid soap bubbles between two transparent layers of plastic sheeting (see page 211). Both systems can be constructed easily, individually, and serve small-scale farming operations.

Temporary Solutions

In many locations where permanent landscaping may not be possible, temporary installations can be used. Sometimes planting systems must be designed to allow regular relocation, and some production or support equipment is designed specifically to be moved around. A variety of flexible containers such as bags and sacks have been used all around the world, from Nairobi, Kenya, where they compensate for land shortages, to San Francisco, where strikingly similar planters formed an art installation by Topher Delaney (see page 217). Another example is the mobile chicken coop, like those used at the Great Kids Farm in Baltimore and Nuestras Raíces Urban Farm in Holyoke, Massachusetts, to allow chickens to be moved around in a secure and easy way while enriching the soil with their organic fertilizer.

Students at the University of Toronto, led by Adrian Blackwell, designed and constructed what they dubbed the Molecular Kitchen, a series of multipurpose carts for preparing and cooking food built primarily from old bicycle parts that can be easily moved around to community gardens and other neighborhood gathering sites, then collapsed when not in use.

Food production can be integrated in these or similar ways with minimal cost, offering significant potential benefit for budding growers in dense or precarious contexts. Some temporary projects such as the proposal for Leadenhall City Farm in London's financial district rely on these types of removable technologies (see page 100).

Crop Intensification and Integration

In dense urban environments where space is at a premium and agriculture competes with other land uses, the focus is often on ways to intensify food production to make it viable for a small space. Some successful proposals integrate food production with building systems. Designers at NY Sun Works have proposed a system of hydroponic trays on a vertical conveyor-belt system mounted between each wall of a double-glazed building facade. This Vertically Integrated Greenhouse (VIG) creates a unique experience for visitors and tenants and turns previously wasted space to a productive use (see page 212).

Stackable or nesting containers also allow vertical cultivation of crops when they are designed to ensure sufficient sunlight penetration and suitable exposure to rain or an irrigation system. Students at Columbia University developed a stackable, concrete, rigid container system called Amphorae based loosely on

the shape of the classical vessels of the same name. The structure is sculptural as well as functional and can be configured in several ways to fit the space at hand (see page 221).

Strategies that integrate crop production include simple, roof-mounted greenhouses such as the one at Maison Productive House (see page 126). It acts as a passive solar collector and a growing space, and benefits from rising waste heat from the building below, making for better growing conditions in the Canadian climate.

The transformation of vertical surfaces into productive space is also considered an integrated system. Many hanging systems have been developed to affix directly to walls or facades. Green wall systems based on modular components most often consist of rigid metal or plastic containers that hold soil for plants, and include built-in irrigation systems. They can be used individually or in multiples to create entire green walls. Flexible, lightweight fabric containers like those made by the Woolly Pocket company can be hung on vertical surfaces and folded flat when not in use (see page 230).

Irrigation

The availability of clean water supplies is an increasing concern in many parts of the world, and one that directly impacts food production. Urban agriculture has to compete for water with industry and household use. Moreover, food production in urban areas often relies on treated mains water for irrigation and other purposes—a waste of an increasingly costly resource. Many waste-water sources in urban areas, however, are suitable for irrigation, so there is enormous potential to develop synergies between food production and water systems.

Municipalities frequently look at policies that will reduce the need for water infrastructure. Rainwater harvesting may be the most obvious and common practice in that helps achieve this goal, and it establishes a direct relationship between cultivation and its surrounding built environment. Another strategy sometimes adopted by municipalities is to encourage the construction of green roofs since they naturally collect water and significantly reduce water run-off. In urban areas, many hard surfaces, including roofs, collect water that has to be disposed of through municipal infrastructure, but when it can be diverted for irrigation purposes it benefits food production and reduces water disposal costs. Many private and community gardens already integrate rainwater collection and storage systems, and some are celebrated as features. Several examples in this chapter feature rainwater-capture cisterns and graywater filtration and irrigation systems. For example, the blue metal-and-fiberglass funnels at the Curtis "50 Cent" Jackson Community Garden in New York (see page

70) decoratively direct rainwater into large cisterns, where it is stored for irrigation.

Also in New York City, GrowNYC initiated a program in 2001 for expanding rainwater-harvesting capacity at community gardens across the city, providing youth with construction training in the process. Water is diverted from the roofs of on-site structures or adjacent buildings into cisterns for irrigation. GrowNYC has installed over sixty such systems, which collect over 775,000 gallons (3 million liters) of rainwater a year.[2]

In cities facing water shortages, integration of food production and water infrastructure can be critical. For example, in Marka, on the outskirts of Amman, Jordan, a communal garden and nursery created by a women's organization atop Amman's oldest landfill created a water-storage cistern beneath a small building and terrace, strategically placed at the lowest spot on the site. Water from the building and terrace and some of the runoff from the site flows into the cistern. This provides adequate irrigation for the garden even in this arid location.[3]

The examples that follow show the potential to establish linkages between the water and wastewater infrastructure and the development of urban agriculture. Most focus on rainwater, but economical and effective ways to divert and reuse graywater are also beginning to be developed.

BELOW: One of the sixty rainwater catchment systems installed by GrowNYC in community gardens across the five boroughs. OPPOSITE: Exploded view of Rios Clementi Hale Studios's Incredible Edible House, showing how many components that support urban agriculture can be incorporated into a single residential unit.

Making Space for Livestock

Debates about the appropriateness of introducing livestock into a city are often what brings urban agriculture to the attention of the public, and the integration of livestock into urban areas remains a contentious issue. Many municipalities still have bans or severe limitations on keeping chickens, bees, or other animals. Urban beekeeping is thriving nevertheless, and has been important to maintain the population given the ravages caused by Colony Collapse Syndrome in rural areas. Urban aquaculture is also growing quickly in response to the devastation caused by commercial fishing practices.

Designers have proposed many components for livestock that make the idea palatable and even fun to the general populace. Urban chicken coops such as the Eglu by the British design company Omlet, and enormous beehives proposed for Detroit by Erika Mayr and Stéphane Orsolini are creative examples that embrace urban conditions and propose solutions distinctive and appropriate to city locations.

Technologies such as aquaponics—which integrates hydroponic cultivation and aquaculture—is an inherently intensive system suited to interiors. Incorporating these systems in residences means adapting traditional house programs to accommodate their needs, such as large tanks of water. The Edible Terrace proposal (see page 136) provides an example of how a single-family dwelling could accommodate this type of closed-loop system.

Conclusion

These examples illustrate the potential for creative components to address urban food production, community engagement, and environmental responsibility. In turn, they create unique urban spaces. The design community is becoming increasingly interested and invested in exploiting emerging technologies to this end. The growBot garden project, for example, brings together designers, artists, farmers, and other food producers to consider how robotics can be used in support of local, small-scale agriculture. Initiatives like these lead to unique solutions and the design of components that are appropriate for urban locations.

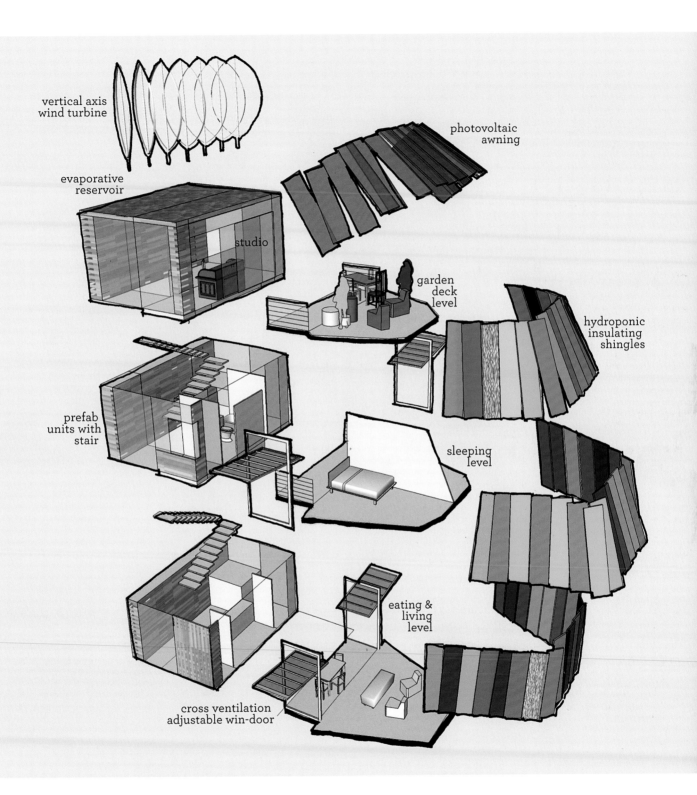

vertical axis
wind turbine

photovoltaic
awning

evaporative
reservoir

studio

garden
deck
level

hydroponic
insulating
shingles

prefab
units with
stair

sleeping
level

eating &
living
level

cross ventilation
adjustable win-door

COMPOSTERS

The practice of recycling organic wastes was historically conducted everywhere, including in cities. In the modern era, however, a shift in practice occurred, and cities began disposing of undesirable byproducts by removing them to an outlying landfill. The benefits of composting have been rediscovered, though, and the practice is now starting to make a remarkable comeback in cities across North America and elsewhere. While industrial-scale composting of municipal waste, which will not be analyzed here, forms the backbone of this revival, composting in backyards, in community gardens, in urban farms, and at other domestic and communal locations has also become commonplace again. Mid-scale neighborhood composting holds great potential for urban gardeners, and indoor systems—such as Growing Power in Milwaukee—as well as outdoor systems—such as the Centre de Compostage Communautaire Tourne-Sol, placed in a visible yet sheltered location in a park in Montreal.

Composting systems require conscientious design that considers materials, functionality, placement, and access. Composters in large communal spaces are typically constructed of ordinary materials like wood, mesh, and nails, and are usually hidden in the least conspicuous part of a garden. Such composting enclosures have gained a reputation for being unattractive, messy, and even hazardous because they are often improvised afterthoughts that are strictly utilitarian. Happily, examples of composting enclosures that are built to be clean, well constructed, and attractive are becoming more common—artists have sometimes been involved in the process of creating them. At the same time, household composting has also become relatively common among homeowners. Most small-scale bins are rectangular or cylindrical, are made of black or dark green plastic, and sit vertically on the ground, meaning that they fit easily and aesthetically into suburban backyards.

Recent innovations also cater to composting within an urban environment, and are designed particularly for urban dwellers without access to a backyard or a communal growing space. These can be used indoors or outdoors and take up relatively little space.

RIGHT: Diagram of
the NatureMill Pro XE
Compost Bin.
FAR RIGHT: An
example of a standing,
roll-type composter
by Gardener's Supply
Company.
BELOW: The Tournesol
Composting Centre in
Montreal, which is open
to the community.
OPPOSITE, LEFT:
Lee Valley Hardware's
plastic rolling-type
composter.
OPPOSITE, RIGHT:
The NatureMill Pro XE
Compost Bin installed
in a residential kitchen
cabinet.

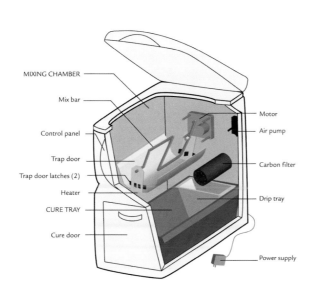

MIXING CHAMBER

Mix bar

Control panel

Trap door

Trap door latches (2)

Heater

CURE TRAY

Cure door

Motor

Air pump

Carbon filter

Drip tray

Power supply

A number of commercially available compost collection bins have been created recently with an aim of making gathering food waste in the kitchen less smelly and more aesthetically pleasing. Countertop and cabinet-door-mounted models are available in a variety of materials, from stainless steel to bamboo to plastic pails with charcoal filters that eliminate odors. After collection, the organics can then be transferred into a larger composting bin.

Numerous variants of these have been designed. More sophisticated models sit sideways on a base that facilitates rotation. Others are compact enough to be placed on a balcony or in a small yard. One model, the Rolling Composter by Lee Valley, has a hatch on the side that opens for easy addition of organic material and removal of compost. Its unique design also collects the "compost tea" that drains from its base, which is an excellent fertilizer and has been shown to discourage pests and diseases. In this unit, compost is created in as little as four weeks because the mix is highly aerated.[4]

Composting at home is also being made easier by a new generation of composters designed to speed up the process. Some add agents that increase the pace of natural composting. Others use electricity to heat the compost slightly, accelerating decomposition. Typically these containers have two chambers: one where food scraps are deposited and composted through mixing, airflow, heat, and moisture; and a second, where finished compost cures the new compost until it is ready or needed. A fan draws air into the unit, providing oxygen to the cultures, while a carbon filter absorbs any lingering odors.[5]

Another way to compost indoors is through vermicomposting, which converts household garbage rapidly into nutrient-laden fertilizer by introducing worms to break down organic matter. The worms shed black castings that can be used as a nitrogen-rich natural fertilizer. While vermicomposting is generally conducted in ordinary containers that can be placed indoors or outdoors, new concepts for indoor units allow them to fit neatly under a sink or in a closet.[6] A number of innovative vermicomposting solutions have been developed in recent years for people who do not like to handle the worms, and for families who produce larger volumes of food scraps than classic composters can accommodate. Some systems like Canada's Worm Inn,[7] are based on a system of hanging bags made of breathable fabric. Others consist of stacking trays that are clean and simple to use;[8] users simply stack new trays on top of trays that contain ready-to-use castings; worms are attracted to the fresh organic matter and make their way up to the new food source through holes in the bottom of each tray.

These examples illustrate the creativity that is being applied by today's designers to ancient techniques for turning domestic wastes from a problem into a resource for the home. Much room remains, however, for further design innovation to be brought to domestic and community composting, from contexts as varied as cheap and replicable neighborhood composters to bins that are designed for the unique constraints imposed by rooftop gardening.

TOP: Section of a simple, bag-based worm composting system by Worm Inn.
ABOVE: The anaerobic digester at Growing Power's Community Food Center in Milwaukee, Wisconsin.

LIVESTOCK SHELTERS

In addition to increased interest in cultivating a variety of crops, city dwellers are also beginning to investigate raising birds, fish, and keeping bees. Restrictive laws are the biggest barrier to keeping certain livestock in some municipalities, but increasingly, these laws are being reexamined. New York City for example has always allowed residents to keep chickens, but recently has changed the law to permit beekeeping as well. The response was overwhelming, with urban farmers and restaurant-rooftop gardeners installing new hives. The right to raise some form of livestock has become a hot political issue in many cities. In Toronto, where chickens are banned, one politician heartily cast his support for a repeal of the law against it saying, "Whether people are raising the chickens to make ends meet, embrace their culture or just have access to fresh food, the city should let people keep the fowl."[9]

Some people raise chickens for fresh eggs; others enjoy raising their own food and want to share the experience with their children and friends; still others see this as a way to increase food security. Industrial designers are responding with creative new products from build-it-yourself kits to ready-to-use prefabricated shelters, including hives, chicken coops, and rabbit hutches that embrace urban conditions and propose design solutions that are distinctive and appropriate to urban settings. Many people, of course, choose not to buy ready-made products but to build solutions of their own design.

Urban henraising has long suffered from a jerry-rigged look and poor construction, which give it a bad name. In New York City, various support programs assist in the construction of coops. The largest program is run by Just Food, an established food advocacy organization. Its chicken program has enabled the erection of a number of coops at community gardens. By bringing skilled labor to interested parties, it has enabled egg production in gardens where members did not feel equipped to construct henhouses themselves.

In Brooklyn, New York, the not-for-profit bk farmyards educates youth on gardening and provides healthy, locally grown food for the borough's food insecure. The founder and chief farmer, Stacey Murphy, and her staff and volunteers accomplish this by using backyards and schoolyards as productive fields. They recently raised a wooden house in a schoolyard that can house fifty hens. This sizable coop is large enough for people to stand in comfortably, and includes an overhanging roof for sun shading. The 100-square-foot (10-square-meter) structure features nests for laying and a completely encased run with hardware cloth buried 12 inches (30 centimeters) below grade to protect the chickens from rats and mice. The coop is raised on stilts to protect the birds from other predators. The eggs supply a Community Supported Agriculture group (CSA) in the community, and children visit the structure to learn about the function of farmyards.

Design firm Omlet responded to the demand for more compact chicken houses with a prefabricated coop, the Eglu, conceived specifically for residential gardens. These distinctive, igloo-shaped, colorful plastic shelters are noteworthy. The coop includes a clip-on metal-mesh fence to allow protected, free-range movement for up to four hens, ducks, or quail. Lightweight construction makes the shell easy to move and removable floor trays and clip-on food dishes make the rearing process simple for less-experienced owners. The Eglu's designers—architecture school graduates—have continued to invent amusing yet sensible, easy-to-use, molded plastic livestock homes including larger chicken houses and a rabbit and guinea pig hutch. The Omlet rabbit hutch is large enough for two rabbits or a few guinea pigs and, like the Eglu, provides a large, secure wire pen so the animals can move around freely.

As with urban poultry, urban beekeeping has a long history. People from many cultures built hives from mud and clay but they—along with the bees inside—were usually destroyed during harvest. Hives with removable parts last many seasons, ensure continuous production, and foster rather than disrupt colonies. Movable-comb hives, where the bee creates a honeycomb on a frame that slips in and out of the hive from the side or top, are the most popular today.

The Beehaus, another product by Omlet sized for the private home, is a triple-insulated shell that maintains the required 95 degrees Fahrenheit (35 degrees Celsius) needed for bees to thrive in both winter and summer. It is modeled loosely after a hive designed by the British engineer Robin Dartington, whose Dartington Long Deep Hives, created specifically for London rooftops, can house two colonies, thanks to an entry on each end. The Beehaus entrances are wasp-proof and designed for easy access. The structure, mounted on legs, puts the hive at a convenient working height.[10]

A 2008 competition entry by Erika Mayr and Stephanie Orsolini for the Holcim Awards[11] conceives a virtual city for bees. A proposed solution to dwindling bee colonies in Detroit, Michigan, this project proposes supplying strategically positioned, oversized beehives to buffer the industrial zones of Detroit from the residential zones and create temporary apiaries and nectar corridors. It was intended to initiate a local beekeeping industry and offer a pollination service for nearby fruit and vegetable farms. The hives are designed to suit local climatic conditions and to raise the bees above the streets, away from people. This project is significant for its design and its vision for large-scale beekeeping in cities.

With compact, easy-to-use, lightweight, and clean modules, the design community is helping simplify the process of raising livestock for the novice city-dweller. Designers are also creating larger, more ambitious shelters for schoolyards and other larger urban farming enterprises to address food needs and quality on a larger scale. The creative solutions resulting from both, mixed with enthusiasm and knowledge, are starting to make substantial contributions to urban food security.

ABOVE: Chicken coop construction at bk farmyards in Brooklyn, New York.
PREVIOUS PAGE: The Beehaus, a modern beehive developed by Omlet in use on a London roof.

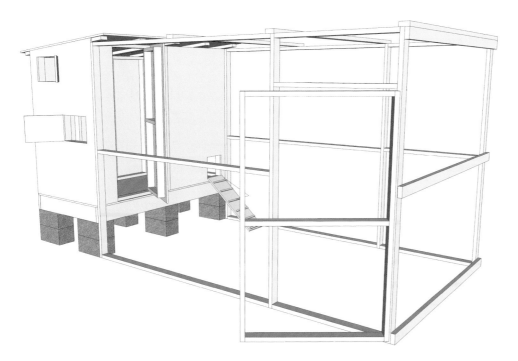

TOP: Rendering for the basic, low-cost coop designed by Just Food, constructed at the Hattie Carthan Community Garden in the neighborhood of Bedford-Stuyvesant, Brooklyn, New York.

ABOVE: Two versions of mobile chicken coops in use at Nuestras Raíces urban farm in Holyoke, Massachusetts, above, and at Great Kids Farm, operated by the Baltimore City Public School System, below.

ABOVE RIGHT: The Eglu, a portable chicken coop designed by Omlet in the United Kingdom, installed in a suburban backyard.

HYDROPONIC AND AQUAPONIC SYSTEMS

Seeds germinated.

Hydroponics and aquaponics are technologies that have the potential to revolutionize the way in which food production can be integrated into architecture by transforming traditionally inert surfaces into productive spaces. The recent appearance of many designs proposed and implemented for green roofs, living walls, and other vegetated surfaces using these technologies is creating excitement about the potential for this ancient technology being reexamined today to grow food in urban areas.

Hydroponics refers to a wide variety of cultivation techniques that grow plants without using soil as a nutrient source, although many hydroponic techniques use a variety of media for structural support. The term hydroponics was first coined in 1937 by Professor William Frederick Gericke of the University of California at Berkeley, who was actively promoting variations of the technology for use in commercial agriculture. Gericke and others recognized that soil is not in itself necessary for plant growth provided that essential nutrients and water can be supplied in another manner.[12]

There are many types of hydroponic systems in use today, from homemade window box planters to multimillion-dollar agricultural operations. The challenge is to find a suitable balance of soluble nutrients for the plant and allowing them to pass over the roots without swamping the plant altogether. The simplest hydroponic systems consist of containers in which plants sit directly in a nutrient solution shallow enough to allow the roots access to oxygen or with a mechanical means of aeration. Nutrient and water levels are controlled manually or automatically. In more energy-intensive

arrangements known as continuous-flow systems, pumps move nutrient solution over plant roots. Many of these use substrates as structural support for plant roots, which also help retain some of the nutrients supplied.

In commercial systems, nutrient concentrations, water supply, and the recycling of the nutrient solutions are monitored and controlled automatically. These systems are both elegant and attractive to those seeking the aesthetic or productive value of vegetation without the need for intensive human labour but raise questions about long-term resilience, for few commercial hydroponic systems can function without a continuous supply of electricity.

Hydroponic systems generally require less water than soil-based cultivation

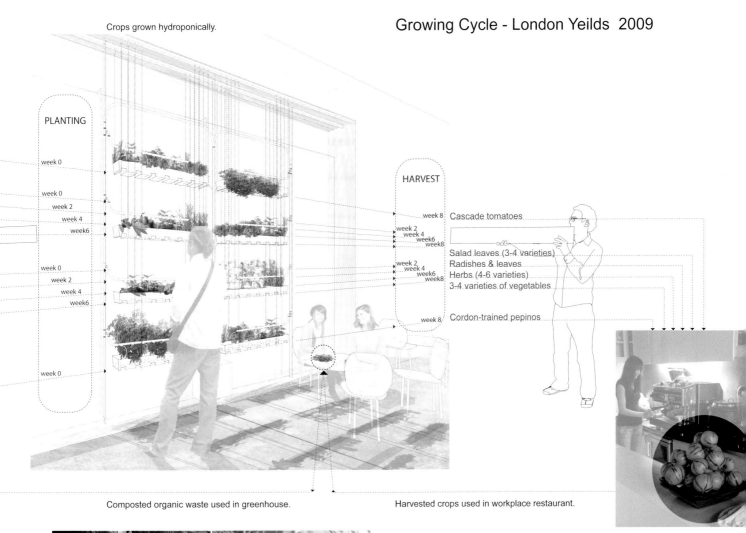

Growing Cycle - London Yeilds 2009

Crops grown hydroponically.

PLANTING

week 0

week 0
week 2
week 4
week6

week 0
week 2
week 4
week6

week 0

HARVEST

week 8   Cascade tomatoes

week 2
week 4
week6
week8   Salad leaves (3-4 varieties)

week 2   Radishes & leaves
week 4
week6   Herbs (4-6 varieties)
week8   3-4 varieties of vegetables

week 8   Cordon-trained pepinos

Composted organic waste used in greenhouse.

Harvested crops used in workplace restaurant.

ABOVE: Process for Bohn & Viljoen's Urban Agriculture Curtain, showing estimated time from planting to harvest.
LEFT: The Urban Agriculture Curtain installed in the London Design Centre.
OPPOSITE: Greens and herbs grown in a hydroponic system at Public School 333 in New York City.

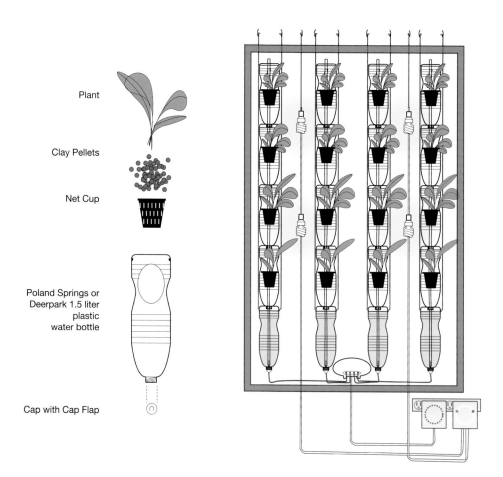

Plant

Clay Pellets

Net Cup

Poland Springs or
Deerpark 1.5 liter
plastic
water bottle

Cap with Cap Flap

ABOVE AND RIGHT: A
hydroponic system designed for
residential use by WindowFarms
based on recycled beverage
bottles and a simple pump.

systems, since much of the water can be continuously recycled. Pollutants in the form of fertilizers or agricultural chemicals also need not be released into surrounding ecosystems or urban areas. The weight of hydroponic systems can often be considerably less than traditional cultivation plots, making them easier to integrate into buildings.

Aquaponic technology is an extension of hydroponics but with animals, usually fish, added into the cycle of food production. In this way, a single system produces a more varied range of food and high yields. Aquaponics is based on age-old knowledge about the waste of domesticated animals containing many of the nutrients required to grow plants. In aquaponic systems, an artificial symbiosis is maintained between fish—grown for consumption—and crops. Certain species of fish like tilapia and yellow perch adapt well to fish farms, grow quickly, and are easily maintained. However, their effluent makes enclosed tanks toxic very quickly. If their water is applied to hydroponically grown crops, however, it supplies valuable nutrients and it is gradually filtered and purified by the crops, allowing it to be reintroduced to tanks to begin the cycle again. A fully closed loop is possible if worms are fed the waste parts of the cultivated plants, their compost is spread over plants, and worms themselves are fed to the fish. Many variations on this process are possible but finding a healthy balance for both fish and plants can be challenging. As a result, aquaponic systems have been introduced slowly to commercial ventures.

Aeroponics is a further development in hydroponic technology, designed to address the problem of aeration of nutrient solutions in hydroponics. Aerosols, or mists, of nutrient solutions are applied to the roots of the plants being cultivated. This system provides optimal aeration of plant roots, uses significantly less water than hydroponic systems, and provides significant increases in biomass yields.

These three technologies are particularly suited to bringing agriculture and biodiversity into dense, urban areas, sometimes resulting in lush, vertical gardens and living walls on facades. Living-wall systems using hydroponic technology are also beneficial to interior air quality. The resulting gradual increase in urban biodiversity in cities may therefore have far-reaching positive consequences on the health of cities.

BrightFarm Systems,[13] the commercial arm of NY Sun Works (see pages 86 and 174), is engaged in a number of projects for commercial developers. Another offshoot of NY Sun Works, Gotham Greens,[14] aims to be New York's first commercial-scale hydroponic greenhouse farm. It will consist of a number of rooftop greenhouses growing pesticide-free fruits, vegetables, and herbs, and will also provide produce to New York's restaurant and catering industries.

The Happy Shrimp Farm[15] in an industrial area of Rotterdam, The Netherlands, is an example of aquaponic technology adapted for shrimp and vegetables. Waste heat from a nearby power station is used to warm the farm's basins to a constant 85 degrees Fahrenheit (30 degrees Celsius), allowing cultivation of Pacific White shrimp for the local market. Waste water containing excess feed and nitrogen-rich shrimp manure is fed into a saltwater vegetable nursery located above the shrimp basins to produce sea lavender and glasswort. The consistent climate conditions allow year-round vegetable production, and the reuse of waste heat and water improved the efficiency of the facility. The locally sold shrimp replace tropical imports that cause various environmental concerns.

Sweet Water Organics[16] in Milwaukee, Wisconsin, has transformed an abandoned industrial building into an aquaponic system that simulates a wetland, featuring fish waste as natural fertilizer for growing plants, which in turn filter the water. It supplies the local community with vegetables such as lettuce and basil, watercress, tomatoes, peppers, chard, and spinach, as well as tilapia and perch. Sweet Water was developed by alumni from Growing Power, a well-known organization promoting urban agriculture, and has advanced its aquaponic techniques while fueling local economic development.

Aiming for a vastly different scale of production is Window Farms, conceived by designers in New York City, which aims to empower food cultivation at home for people who lack a yard.[17] Its open-source hydroponic system made largely from recycled bottles can be hung in urban windows and used to grow small quantities of food. It also sells a complete kit. The Urban Agriculture Curtain (see the Greenhouse Technologies section) uses a similar concept to grow food in windows.

These examples, from simple hanging recycled bottles to complex master plans for entire villages, illustrate the wide scope and potential applications for this technology. Several projects featured in this book contain hydroponic or aquaponic components. These include the Edible Terrace proposal (see page 136), which integrates a fish tank in the basement and growing spaces at ground level within a redesigned row house; Greenhouse Village (see page 41), which considers similar approaches but at the scale of a an entire community; and the Science Barge (see page 86), an educational venue that creates pesticide-free produce in a closed-loop system.

Resilient and sustainable food systems can be implemented in large cities using small-scale components with a mass application. Gradual implementation of artisanal urban food production is a solution to the serious problems of urban dependence on increasingly unsustainable and unstable industrial agriculture.

GREENHOUSE TECHNOLOGIES

The ability of greenhouses to capture solar energy is one of the key principles of passive solar design. Greenhouse technology offers many advantages for integrating food-growing systems directly into buildings conceived principally for other purposes: in addition to potential energy savings, it serves as a buffer zone for habitable spaces, extends the growing season to improve yields, and allows the cultivation of delicate and exotic plants that may otherwise not survive in exposed conditions. Greenhouse technologies can also be combined with strategies such as hydroponics and aquaponics to further increase production.

The popularity of greenhouses used to grow fruits, especially oranges, in Europe from the seventeenth century on stimulated technological developments in heating. The evolution of the greenhouse is one of a gradual lightening of structures, increasing of spans and refining of materials. The Industrial Revolution brought rapid advancements in structural technologies and an increase in the availability of affordable glass, leading to the great glass palaces of nineteenth-century expositions, and also allowing the technology to gradually become more accessible and economically feasible for commercial and domestic uses.

Recent decades have seen technological breakthroughs in glazing systems that reduce energy consumption in the form of heating or cooling to maintain consistent temperatures. Today people grow food under glass in practically all climates. Reduced overall weight and improved thermal performance mean that greenhouse systems can also be more readily integrated into, or on top of, buildings. NY Sun Works, for example, has designed several rooftop and floating greenhouse projects that combine lightweight structures, modern glazing systems, and hydroponic growing techniques to make intensive farming feasible

as well as profitable. NY Sun Works uses waste heat from buildings along with solar heat to keep rooftop greenhouses productive throughout the winter without relying on unsustainable, fossil-fuel-based energy sources.[18]

Double-skin facades created from two layers of glazing separated by a large air space, and with integrated ventilation and solar control, are a relatively new technology beginning to be introduced into buildings with large curtain walls. A research group composed of NY Sun Works, Arup Engineers, and Kiss + Cathcart Architects have designed a system that combines the features of a double-skin facade with a hydroponic greenhouse so that the air space can be used for food production. The system, known as a Vertically Integrated Greenhouse (VIG), is intended for south-facing, unobstructed facades on tall buildings. It consists of two columns of hydroponic planters suspended on cables that cycle slowly between the two layers of glazing. Plants are planted in stages,

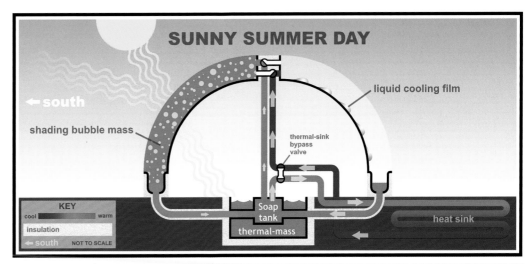

SUNNY SUMMER DAY

← south

shading bubble mass

liquid cooling film

thermal-sink bypass valve

Soap tank

thermal-mass

heat sink

KEY

cool ▬▬▬ warm

insulation

← south   NOT TO SCALE

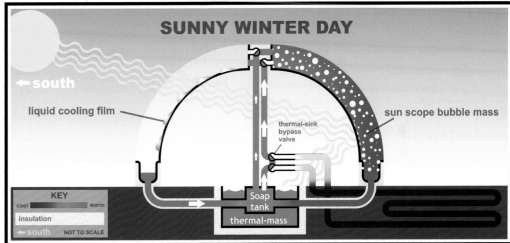

SUNNY WINTER DAY

← south

liquid cooling film

sun scope bubble mass

thermal-sink bypass valve

Soap tank

thermal-mass

KEY

cool ▬▬▬ warm

insulation

← south   NOT TO SCALE

RIGHT: Diagrams of the Solar Bubble greenhouse concept, showing how soap bubbles harness the sun's energy in different seasons.
BELOW: The greenhouse at Roberta's Pizza in Brooklyn, New York, fabricated entirely of discarded window frames.
OPPOSITE: Two views of the Green Market proposal by Kiss + Cathcart, showing a VIG system designed to line the glass curtain wall and ceiling with plants.

and ideally, a single cycle corresponds to the length of time it takes for a given plant to bear a crop. In this way, all harvesting could be done from a single point at street level. The hydroponic planters can also be tilted like Venetian blinds to control the flow of sunlight to the building's interior. More than simply a thermal buffer zone, the VIG absorbs excess solar radiation and wraps occupants in a lush green envelope. VIGs can be incorporated into new projects as well as existing buildings.

GreenMarket is a proposal for a prototype building utilizing VIGs in Masdar City, United Arab Emirates. It was designed by Bright Farm Systems—the commercial partner of NY Sun Works—and Kiss + Cathcart Architects for use in a large-span hall that can accommodate a variety of programming. Most of the exterior walls would consist of VIG systems and provide the building's owners with excess income from selling the crop. The proposal also suggests that a similar system to a VIG could be incorporated between the tall trusses that form the roof structure. As such, the building would be completely enveloped in a high-performance, productive double-skin facade. The roof would be formed of a material to admit ample daylight while the VIG shaded the interior and produced food in a single gesture.

For smaller-scale projects, Bohn and Viljoen Architects—working with Hadlow College—have designed a compact, high-yield, low-maintenance, hydroponic growing system called the Urban Agriculture Curtain. It can be installed behind existing glass facades or on small balconies and is suitable for situations

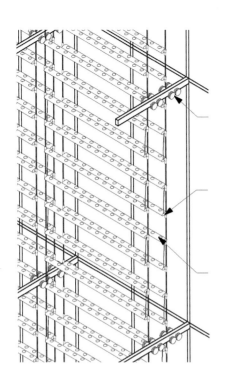

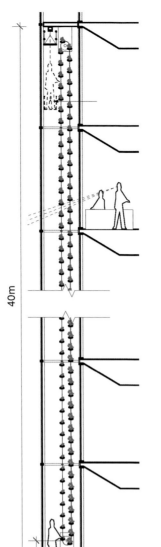

ABOVE, RIGHT, AND BELOW: Diagrams and a rendering of a Vertically Integrated Greenhouse, designed to be installed between the panes of glass that compose a high-rise office building's curtain wall.
OPPOSITE: Rendering of the Aquapod, a portable, self-contained greenhouse system.

where high yields are required and minimum gardening time is available, such as in apartments, offices, or cafés. The installation uses standard building components customized to the available space. Eight planting trays are hung on an off-the-shelf cable system and connected to pipes that supply them with nutrient-rich water, which is stored in a nearby tank. The planting trays are turned once a week to expose plants equally to the sun, but the water tank must be checked on a more frequent basis. Food-growing spaces do not consume floor space and plants provide a degree of privacy and shading. The installation is plumbed in a similar way to a washing machine and therefore could also be fitted for residential apartments. The WindowFarm, a similar concept (see page 208), uses recycled bottles to create small, hydroponic, indoor grow-ing areas in front of traditional window spaces for food production.

A variety of plastic materials also offer interesting opportunities and have been used widely in commercial greenhouse applications. The Solar Bubble Greenhouse consists of two layers of plastic sheeting suspended over a double-hoop structure through which liquid soap bubbles, generated at the top of the structure, can flow.

Depending on the time of day, year, or weather, the bubbles serve either as an insulating layer on cold days or a screen that reduces solar radiation. The system is also capable of drawing excess heat out of the greenhouse through a process similar to expiration in plants. Daytime heat gain is stored in thermal mass components within the space and released slowly through the night. The solar bubble system, while dependent on a constant energy supply itself, can significantly increase the greenhouse's efficiency and reduce its overall energy consumption while using a very light frame.

Solar bubble technology is being integrated with a variety of related sustainable strategies including a window farm, hydroponics, a renewable energy system, and worm composting in the conceptual Aquapod project. This innovative greenhouse design provides educational facilities centered on food production, alternative energy, and green building. The Aquapod uses the lightweight solar bubble system to admit sunlight while conserving heat on the interior to create optimal growing conditions. The Aquapod produces the greatest possible yield on a small footprint while efficiently using water and energy. Its design and analytical team has applied for funding for the construction of a prototype.

Using simpler technologies it is possible to creatively reuse old window frames to produce cold frames or even small-scale greenhouses. In fact building full-size greenhouses from recycled materials can be an attractive undertaking both aesthetically and environmentally speaking. Roberta's Pizza, a unique restaurant and farm located in Brooklyn, New York, has built two greenhouses directly behind the restaurant in order to increase the freshness of their ingredients and reduce the distance that they traveled. The first, a simple hoop structure with a plastic skin, was placed atop an old shipping container adapted into offices for a food-focused radio station partly heated with excess heat from the restaurant's kitchen. When the first proved successful for the restaurant's needs, windows from local industrial buildings destined to be discarded were gathered and pieced together on a timber frame to form the walls of a second greenhouse. The lightweight roof was constructed from corrugated polycarbonate sheets. The result is a highly functional, albeit rough-looking structure that matches the individual, earthy style of the restaurant. At a larger scale, entire buildings have found new life as special types of greenhouses: in Canada's Northwest Territories, a hockey arena was converted to a community greenhouse (see page 67) and Toronto employed adaptive reuse to turn a streetcar repair barn into a technologically advanced greenhouse (see page 74).

Greenhouse systems can range from the very simple, created from repurposed, found, or used components, to very sophisticated, state-of-the-art technologies. Designers of late have used their form to create ingenious spaces for urban agriculture that often contribute greatly to the built environment. New building forms and a new architectural language is beginning to evolve out of the integration of greenhouse systems with other building components and systems.

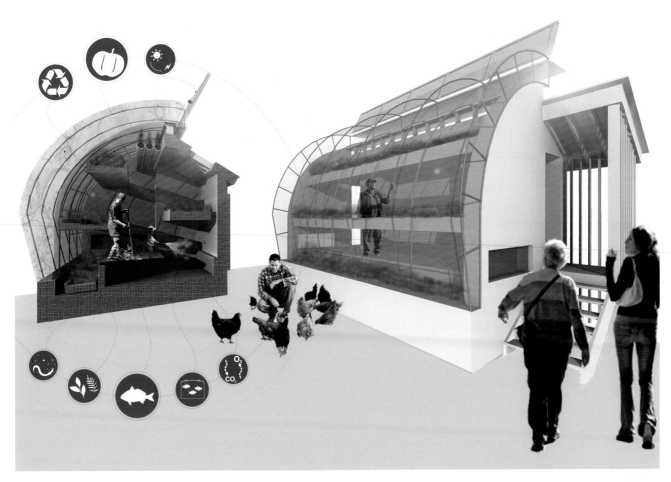

RAISED BEDS

One type of container for cultivation warrants particular attention for its suitability to the urban context: the raised bed. Anywhere the ground is not accessible, the soil is not safe, or remediation is not feasible, raised beds may be introduced as a viable method for cultivating an extensive area. Raised beds are often the method of choice for agriculture on "brownfields" due to concerns about land contamination. Philadelphia's GreensGrow, for example, has operated for years on former industrial lands solely by growing plants without digging into the ground; as it evolved, production techniques were diversified through the addition of more expensive and sophisticated methods—but the use of a variety of containers, both movable and fixed, such as raised beds, remains at the core of its operations. At South East False Creek, raised beds are used for similar reasons.

Raised beds can be assembled simply from ordinary materials, typically wood, which anyone with basic tools and construction skills can create and put together. The simple forms, inexpensive materials, and low maintenance needs that are the hallmarks of raised beds—and which make them so practical—unfortunately also means that they can look ramshackle and dilapidated if not well planned, giving urban agriculture a bad reputation as a generator of a

shoddy, uncared-for looking landscape. In reaction, some recent urban agriculture projects have given particular attention to finding creative solutions for the design of raised beds, with playfulness being the key principle.

For example, the rectangular raised beds adjacent to the farmers' market at Billings Forge, an adaptive reuse of a historic factory in Hartford, Connecticut, are simple but colorful, well-constructed, and regularly maintained. Similarly

constructed wooden raised beds were used at the Growing Joy Community Garden in Detroit, and are painted red, white, and blue to provide a pop of patriotic color.

In other instances, location and form drive the inspiration for the playfulness. An outstanding example is in Birmingham in the United Kingdom, at a new container-based growing scheme on municipally owned land.[19] The residents who will be the garden's users specifically requested that the containers vary in size—since the area slated for cultivation was not the same for each household—as well as in layout. This bestows an intentionally random look on the garden, in reaction to the rigid raised-bed look of formal gardens, which are often laid out in straight rows.

In other cases, the shape of the raised bed itself is transformed, sometimes to striking effect. A creative adaptation of typical rectilinear, wooden containers can be found in Vancouver, British Columbia,

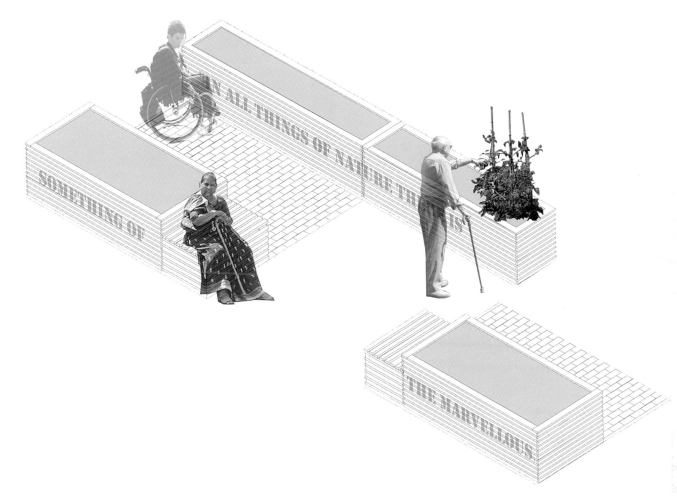

TOP: Raised planters designed by What if: projects for one of the Vacant Lot sites in London are intended to be accessible for the housing estate's elderly and disabled population.
ABOVE: Raised beds at the NutriCentre project in Montreal were designed to allow wheelchair access.
OPPOSITE: Hexagonal beds installed at the My Own Backyard, or MOBY, community garden in Vancouver, British Columbia.

at the My Own Backyard (MOBY) community garden. MOBY was initiated in 2003 to transform a vacant lot underneath the elevated tracks of Vancouver's rapid-transit trains that had been overflowing with garbage into a garden with over forty plots and a corn-cob construction shed, playground, and green space.[20] The beds took just two days to build, and were constructed by community volunteers. Irrigation lines were installed, allotting one tap to every six beds.[21] This project stands out visually for its hexagonal beds arranged to look like honeycombs. MOBY's garden organizers chose this shape for its symbolic value—the beehive represents the community—as well as to deliberately reinvent the typical linear shapes and placements associated with raised beds in general.

London-based design and advocacy firm What if: projects, known for their inventive planter designs, now also offers an inventive raised-bed kit. Its development began when the team received funding to create twenty food-growing spaces for public housing residents as part of an extension of their Vacant Lot initiative.[22] To bypass concerns for soil contamination and to facilitate flexibility, a modular approach was applied to all food-growing areas. Bulk bags were employed on hard surfaces and raised beds on grassy surfaces. The latter features an innovative system of flat-pack, lightweight timber planters that are fabricated by What If and can be easily transported and assembled on site. The adoption of the flat-pack approach means the planters can be easily installed near seating areas, water towers, steps, fences, and storage sheds. The modularity allows easy adaptation to many contexts with different needs, while ensuring a clean, unified-looking design across the entire site.

Some of the new Vacant Lot sites were designed specifically for groups with limited physical abilities. Raised beds can be used to create gardens with greater accessibility by bringing the plants closer to people, and are increasingly important given the mushrooming of the world's elderly populations who wish to continue to work the soil. A number of good examples of gardens designed for this purpose exist, such as the Malthouse Garden in Smethwick, in the United Kingdom, which was developed by the organization Ideal for All with the consultation of people with a variety of disabilities. McGill University's Minimum Cost Housing Group similarly partnered with Nutri-Centre, a community development organization, in Lasalle, one of Montreal's poorest boroughs. This involved a redesign and upgrade for a collective garden to enable greater accessibility for the mobility-impaired. Spacing between the new raised beds, as well as their proportions, were designed with the help of residents

so that wheelchairs can pass easily through the garden, and are notched so that wheelchairs can gain access to plants without having to overreach.

Raised beds are being reinvented in multiple ways; they are no longer simply an easy means of creating spaces for cultivation without dealing with ground conditions. Instead they are becoming design objects in their own right. The reuse of materials, accessibility, modularity, and aesthetics they can offer, are being fully explored. The raised bed illustrates well how many traditional components of urban agriculture are being adapted to enhance their role in urban contexts.

ABOVE: Contaminated waste spaces in Detroit are farmed with the aid of colorful wooden raised beds, shown during an annual tour of its urban agriculture operations.

SOFT PLANTERS

In many locations where permanent landscaping may not be possible or desired, temporary planters constructed of collapsible materials are often introduced. A variety of creative soft planters have been used all around the world for different purposes, from simple bags that expand available arable land in Kenya and Great Britain, to specially sown bags that became art installations in the United States.

In developing countries high food prices and low incomes make food sovereignty a major issue. Locally produced food can be an important source of nutrition for their people. A great deal of ingenuity is often used in poorer places to create food-growing opportunities using minimal resources. In the densely populated slums of Nairobi, Kenya, for example, a shortage open land has led many residents to adopt a simple and innovative growing method: "a farm in a sack." The Italian NGO Cooperazione Internazionale (COOPI) claims that, with appropriate soil content, up to forty seedlings can be planted in each woven sack. Seedlings grow out of holes punched in the sides of the filled bag, using maximum surface area. This system works for vegetables like kale, spinach, squash, spring onions, and sweet peppers, and it can feed one household for at least two months. Space limitations have been overcome by placing them on neglected areas such as roofs, doorsteps, and porches. Sacks are also used for the larger community on vacant sites.[23]

A similar concept has been explored in London, United Kingdom, by What If: projects, an art and architecture practice that engages local communities to implement sustainable environments. Vacant and neglected spaces on inner-city public housing land is transformed with removable polypropylene bags. Polypropylene is a material used internationally for the wrapping and shipping of goods due to its low cost, flexibility of construction, and durability. These bags are also used in the construction industry to move building materials and waste to and from site. These large, flexible containers can be easily transformed into practical planters for produce. In 2007, What If used polypropylene to create seventy 11-square-foot (1-square-meter) bags with distinctive graphics. These are used by local residents in Shoreditch to grow food on an unused paved area.[24] The bags are each filled with half a ton of soil. An array of fruits and vegetables, ranging from salad greens and berries to root crops and flowers are grown. The garden has become a social space for neighbors, barbeques, sunbathing, and playing. Plot holders of different age groups and cultural backgrounds meet and exchange food, seeds, and gardening advice. This successful experiment is being expanded to other low-income neighborhoods in London.

TOP: A rendering of Gretchen Hooker's Soft Planting System designed for San Francisco's Digging Deeper competition.
ABOVE: "Farm in a Sack" bags used as planters in Kibera, an urban slum outside Nairobi, Kenya.
LEFT: Three of Gretchen Hooker's temporary planting containers made of lightweight, breathable, polyester fabric.

San Francisco–based artist and landscape designer Topher Delaney has also used recycled polypropylene to craft sculptural containers as bags for plant growth.[25] The orange-and-yellow Big Bags are made in Sweden and used for hauling material to and from construction sites. Delaney cut up forty bags provided by the manufacturer to form large, flat sheets of polypropylene, then reshaped them into interesting forms that retained the brand's recognizable graphics. The addition of zippers on some panels made the planters look like huge flowers opening petals to reveal the plants that flourish within.

An entry by Gretchen Hooker for San Francisco's Digging Deeper competition proposes the "Soft Planting System," intended to transform urban spaces into edible landscapes. Her intention is to provide individuals and groups with opportunities to quickly and easily set up urban gardens in diverse opportunistic spaces, which may be temporary. Her lightweight, modular containers are made of polyester fabric woven to create flexible, breathable planters that have pocketlike openings in the sides to increase growing area. The malleable, triangular form allows modules to nestle together to maximize the use of space and can easily be adjusted to suit site-specific conditions. A tensile canopy of weatherproof fabric can be used to provide shade and an open knit trellis can also be added. The system is economical, can be made in a range of sizes from 10 gallons (40 liters) to 50 gallons (200 liters), and can be easily removed and reused or recycled.

Children's wading pools have been reused in several recent projects as planters. These are particularly suitable for rooftop gardens due to their light weight and shallow soil. At the Noble Rot restaurant rooftop garden in Portland, Oregon, Edible Skylines and Urban Agriculture Solutions devised a variation on hydroponics using these small pools. Each was filled with 2 to 4 inches of vermiculite or gravel and covered with soil mixture. The pools are filled with water to the top of the vermiculite layer and roots reach down from the soil into the water layer. This simple technique takes less water, time, and equipment than overhead or drip irrigation.

There are many inventive ways to create simple, flexible planters using discarded or readily available materials. Projects in this book such as the Leadenhall City Farm proposal (see page 100) use such planters in bags to create temporary growing spaces within larger projects. In addition to generally weighing far less than hard planters, soft planters are often easier to put away in the off season, making them especially suited for apartment residents and others with limited storage space. Some designers have seen an economic niche in the creation of soft planters for sale to such populations, as illustrated by BacSac and Woolly Pockets (see page 230). Food production can be integrated into many spaces in this way with minimal cost, offering significant potential benefit for budding growers in dense or challenging situations.

TOP AND ABOVE: Vacant Lot temporary planting bags made of polypropylene, developed by What: if projects in London, installed on an urban waste lot of formerly unused space at a social housing estate.

RIGID CONTAINERS

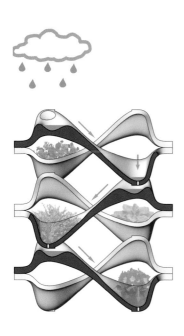

Rigid containers have been used for cultivation since ancient times. They are common wherever the ground is unsuitable for planting, or when people desire to be able to move plants occasionally. Clay pots scattered around a courtyard in a warm Italian town, for example, or window boxes hanging from a high-altitude Swiss chalet illustrate how present the rigid container is in the imagery of urban life, from the smallest villages to the largest cities.

Planters come in many shapes and sizes because they can be fashioned from almost anything. Some are simply reused vessels intended for other purposes originally. At the farm of the Southside Community Land Trust in Providence, Rhode Island, for instance, an abandoned claw-foot bathtub and an old washing machine have both found a new life as planters for herbs. Stacks of used tires are another ever-popular option for low-cost planters. The options for reuse of materials for rigid containers is truly limited only by the inventiveness of the gardener.

Even simple containers involve several basic design decisions: which materials to use and where to source them, how the water can drain, where to place the containers, and sometimes how to make them easy to store or stack. These basic considerations are being incorporated into a host of fresh new designs as they are adapted to different urban conditions and uses.

Some rigid containers are freestanding structures, such as the colorful molded-plastic combined seat/planters known as Portable Gardens that hold herbs and olive trees, and were designed by artist Leonel Moura. These grace the waterfront in Lisbon, Portugal.[26]

Other rigid containers are intended to interlock to create a network of self-watering planters that spread moisture via capillary action, and are designed for use on open-air flat surfaces such as rooftops.[27] Such systems are a hybrid between the full rooftop coverage of planted green roofs and modular, individual containers. The two main advantages of capillary-action systems are that they provide a good environment for growing healthy vegetables while avoiding root rot, and they are low-maintenance.

For the last five years, Alternatives, a leading Montreal-based NGO, has been fostering the participatory cultivation of edible landscapes in cities; to achieve this, a team led by Ismael Hautecoeur developed a planter with a water reservoir

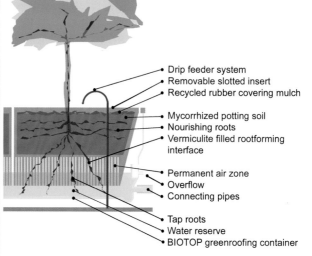

Drip feeder system
Removable slotted insert
Recycled rubber covering mulch

Mycorrhized potting soil
Nourishing roots
Vermiculite filled rootforming interface

Permanent air zone
Overflow
Connecting pipes

Tap roots
Water reserve
BIOTOP greenroofing container

TOP AND ABOVE: The BioTop lightweight container system; it aims to produce three times as much produce as conventional systems while using half the water.
LEFT: The Tower Garden by Future Growing in production on the roof of Bell Book & Candle restaurant in New York City; the system supplies produce used in the 89-seat restaurant below.
PREVIOUS PAGE: The modular, sculptural Amphorae interlocking outdoor plant growing and watering system, developed at Columbia University in New York City.

ABOVE: Herbs growing in EZGro's tower system containers, adapted by BrightFarm Systems, at a Whole Foods grocery store in New Jersey.

for low-maintenance vegetable gardening. This container system (see page 94) adds a false bottom that hiding two water reservoirs to the interior of an ordinary plastic recycling bin. The reservoirs are filled simply through a vertical tube that stands upright in the bin. The team also created a manual explaining how to create a similar planter using any rigid container.[28]

Other capillary-based container systems now available to the public are the BioTop and the EarthBox. Both are patented low-maintenance growing systems that include a standard container, a screen, a water-filling tube, reversible mulch covers, lightweight growing medium, fertilizer, and instructions.[29] EarthBox containers are being used by students in school gardening programs and community gardens in a number of countries through a program called The Growing Connection, operated by the United Nations Food and Agriculture Organization. Students in various locations use the Internet to share lessons, experiences, and horticultural experiments with each other.[30]

Rigid containers can also be stacked vertically, saving space while creating sculptural structures. Multilayered rigid containers tend to be more expensive than other versions and can require sophisticated engineering and construction in order to function properly, however. Creative designs for stacked rigid containers are nevertheless plentiful.[31] One type employs advanced growing techniques, such as hydroponic systems. A second relies on the development of special materials and construction

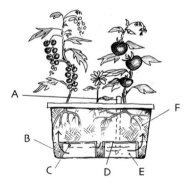

A: Filling tube

B: Submerged strip
(Underground irrigation by capillarity)

C: Overflow hole

D: False bottom

E: Water reservoir

F: Soil mixture

techniques for the form of the structure itself. And the third main type, containers made of ordinary materials, often benefit from low-tech support structures to use space efficiently.

New York City–based BrightFarm Systems designed and installed a demonstration-scale, indoor food-production assembly for a Whole Foods in Milburn, New Jersey, using water-efficient, recirculating hydroponics. The design team delivered an effective growing system for a space with sub-optimal lighting conditions—a room that receives natural light on one side only.[32] A tower system by EzGro made of flower-shaped pots stacked on a central pole was adapted for this context.[33]

The Tower Garden by Future Growing is basically a perforated long tube over a 25-gallon reservoir. Its internal system, powered by solar panels, recirculates water and dispenses nutrients. It is made from food-grade plastic that has been UV stabilized for years of outdoor use. In the summer of 2010, one was installed by a restaurant's owner/chef on its rooftop. The system of 60 short towers contains over 1,000 plants, and the first harvest took place four weeks after installation.

Like these hydroponic tower systems, the Amphorae system is also based

on water flow and is fully stackable. These units are constructed from Ductal concrete composed of a mixture of recycled sand and reconstituted fly ash. They are designed with small pockets for plants, which are irrigated by built-in reservoirs. Amphorae were cast from a reconfigurable mold that can dictate different functions based on program and positioning. The blocks allow vegetables to grow as part of facades or independent structures, making up a wall, an arch, or a canopy.[34]

A simpler product, the Tomato Plant Stand designed by What If: projects from London, England, represents a simple, effective, approach to stacking rigid containers, and relies on pots instead of plots. It uses ordinary flowerpots set upon shelves made from common lumber. Such systems are intentionally simple, flexible, and easy to assemble and relocate.[35]

Rigid containers have evolved from the found object to highly engineered forms; from reused materials to new, industrially generated compounds; from the easy-to-move small containers to the interlocked and stacked assemblies. The range of solutions that build on the rigid-container idea are varied and can suit any location.

LEFT, ABOVE: Recycled plastic bins become rigid containers for plants at Roberta's Pizza in Brooklyn, New York.
TOP, ABOVE: The EarthBox in a greenhouse of the Baltimore City Public School System's Great Kids Farm.
CENTER, ABOVE: Alternatives planters in use on the roof of the Université du Québec in Montreal.
ABOVE: Rigid containers and trained vines at McGill University in Montreal.
LEFT: Unexpected found items such as bathtubs like this one installed at the City Farm of the Southside Community Land Trust in Providence, Rhode Island, can make useful rigid containers.

Planters for olive trees and herbs, designed by Leonel Moura, double as seating near the Cais do Sodré Train Station and Ferry Terminal in Lisbon, Portugal.

HANGING PLANTERS

Food cultivation was for eons literally grounded, dependent on the availability and quality of fertile soil in the fields surrounding areas of human habitation. Yet the history of separating cultivation from the ground is also long and varied, and arose particularly in response to the tight confines of dense urban settlements, where the ground is reserved for many other uses such as buildings, streets, and common areas, or is not to be trusted because of fears about its conditions. Vertical surfaces can be exploited to combine food production with other benefits such as shade, air purification, and aesthetics. The climbing grapevine is perhaps the ultimate example of vertical cultivation that also provides shade and greenery.

Unlike examples of vertical agriculture systems shown in other sections, which sit in or on the ground, in these examples a complete separation of plants from the ground is achieved through hanging objects for cultivation from vertical surfaces. Recent examples of this style of agriculture fall into two main approaches: the use of rigid, modular, systems and the use of flexible fabric. Both systems can be found indoors or out, and can be used at a small scale for individual plantings, or combined to create larger swaths of green facade.

Mobile Edible Wall Units (MEWU) are plant panels mounted on wheels for easy relocation to maximize the plants' exposure to the sun throughout the year, and are configured to fit odd-shaped indoor and outdoor environments.[36] The MEWU system is the result of a partnership between Barthelmes Manufacturing Company Inc.[37] and Green Living Technologies, which contributed design and marketing. The panels are notable among hanging containers for being able to accommodate deeper-rooted plants—tomatoes, cucumbers, bell peppers, leeks, eggplant, lettuce, spinach, and

even baby watermelon have been produced successfully. Manufactured from food-grade stainless steel and high-grade aluminum, the panels feature intersecting slats that create cells for growing media and a patented design that allows for water flow to prevent waste and root migration to prevent root rot. MEWUs have been used for the benefit of the community in Los Angeles, where a series of edible walls helped to feed the homeless. At Discovery High School in The Bronx, New York, teacher Steve Ritz used the edible walls to engage students, increase attendance, and improve grades.[38]

ELT's Easy Green living wall system is another example of vertical productive panels. This modular system is made from reusable, recyclable, polypropylene copolymer. Easy Green panels can be mounted on exterior or interior wall surfaces and are designed to allow water to flow without pulling the soil away with it. Soil retention is achieved through the use of angled cells that retain some water, and notches at the top and bottom of each cell that allow for irrigation, drainage, and aeration.[39]

An alternative approach uses soft materials to create systems of hanging containers that are, essentially, variants on the concept of a hanging sack. When filled, these bags often weigh 2.5 pounds (1 kilogram) or less, making them well-suited to settings with load limits. They often feature built-in moisture barriers to help protect furniture and interiors, making them equally suitable for indoor or outdoor use. These products are designed to be lightweight and foldable, and thus easy to move and store—essential characteristics for the urban user.

BacSac, a permeable garden bag produced in France, is an example of a hanging sack. It was created as a solution to urban roof garden constraints: difficulties with transportation, excessive weight, and lack of attractive containers. BacSacs are constructed from a geotextile and are recyclable. Their double-walled construction maintains the necessary balance between air, soil, and water and is resistant to frost and sun. BacSac is produced in a variety of styles, including a window-box compatible bag, and a saddlebag that can hang over a railing or be draped over an A-frame support,[40] in addition to planters that sit on the ground.

Another example of flexible, breathable and modular gardening bags are offered by Woolly Pockets, which are handmade in Phoenix and Los Angeles. Woolly Pockets

ABOVE: Mobile Edible Wall Units installed in an urban area increase the site's arable surface.
BELOW: A polished version of Green Living Technologies's Mobile Edible Wall Unit.

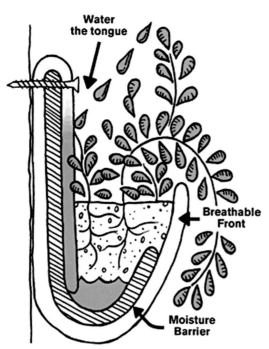

have two main components: a breathable fabric made of recycled plastic bottles that have been industrially felted, and a built-in, impermeable moisture barrier made of 60 percent recycled plastic bottles. The company's Wally model is designed for vertical applications, based on a standard 15-by-24-inch module that is available in single, three-, or five-unit pockets. Bags can be combined to cover an entire wall if desired.[41]

Edible walls are essentially a subset of the green wall concept. Hanging containers detach the growing medium from the ground, and they reinvent the wall as a productive surface. The placement of hanging containers in relation to the sun—even inside a building, if the light conditions are adequate—help to make formerly unused urban spaces productive and bolster health. Hanging containers are an important, simpler alternative to contemporary living-wall systems, which typically use elaborate mechanisms for maintaining greenery on a wall. They recall older techniques of incorporating agriculture into urban spaces such as espaliers, adding historical context. Hanging containers are a key instrument for making all three dimensions of a building—floor, roof, and wall—productive.

NOTES

INTRODUCTION

[1] Many affluent cities still produce significant amounts of food within their built-up area. In the United States, over one-third of the agricultural production *by value* takes place within Metropolitan Statistical Areas. Even in the densest parts of the world, intensive food production takes place to take advantage of the local market. The Dutch *randstad* may be the most famous case of a "green lung" that is vital to a conurbation. As recently as the 1990s, Hong Kong produced 66 percent of the poultry, 16 percent of the pigs, and 45 percent of the vegetables eaten by its citizens, while Singapore grew 25 percent of its own vegetables and is relatively self-sufficient in pork, poultry and eggs. These statistics cited in Jac Smit, Joe Nasr, and Annu Ratta, "Urban Agriculture: Food, Jobs and Sustainable Cities" (sponsored by the United Nations Development Programme, unpublished 2nd edition, 2001). Posted at http://www.jacsmit.com and http://www.metroagalliance.org.

[2] Many different definitions of *urban agriculture* exist. Smit, Nasr, and Ratta define it as:

> An industry that produces, processes, and markets food, fuel, and other outputs, largely in response to the daily demand of consumers within a town, city, or metropolis, on many types of privately and publicly held land and water bodies found throughout intra-urban and peri-urban areas. Typically urban agriculture applies intensive production methods, frequently using and reusing natural resources and urban wastes, to yield a diverse array of land-, water-, and air-based fauna and flora, contributing to the food security, health, livelihood, and environment of the individual, household, and community.

An analysis of approaches to defining urban agriculture can be found in Appendix A of Soonya Quon, "Planning for Urban Agriculture: A Review of Tools and Strategies for Urban Planners" (Cities Feeding People Series Report 28, Ottawa: International Development Research Centre, 1999). Posted at http://www.idrc.ca/uploads/user-S/10739436551Report28completed.doc.

[3] American Planning Association, *Policy Guide on Community and Regional Food Planning*, American Planning Association, May 11, 2007. Posted at http://www.planning.org/policy/guides/adopted/food.htm.

[4] The U.S. Green Building Council's program for neighborhood development (LEED-ND) includes a credit for projects designed with permanent farms and productive gardens.

[5] "The Role of Food and Agriculture in the Design and Planning of Buildings and Cities," held at Ryerson University, May 2–4, 2008. Most presentations can be found at http://www.ryerson.ca/foodsecurity/projects/urbandesign/presentations.html.

[6] The exhibit *Carrot City: Designing for Urban Agriculture* was first shown at Toronto's Design Exchange from late February through April 2009. Versions of the exhibit were later mounted at additional venues in Toronto, including Canada Blooms and the Royal Agricultural Winter Fair. The exhibit has, as of publication, also traveled to Parsons The New School for Design in New York City in fall 2010, to the American Institute of Architects–Rhode Island, and at La Coupole, Parc de la Ligue Arabe, Casablanca, Morocco in April 2011. Many cases included in this book were first researched for and presented at the exhibition.

[7] See http://www.ryerson.ca/carrotcity.

[8] Ebenezer Howard, *Garden Cities of To-Morrow* (Cambridge, Mass.: M.I.T. Press, 1965). First published as *To-Morrow: A Peaceful Path to Real Reform* (London: Swan Sonnenschein, 1898).

[9] Sir Patrick Geddes, *Cities in Evolution* (London: Williams & Norgate, 1915). His Valley section can be found at http://www.transect.org/natural_img.html. For the adaptation of the transect by the Regional Planning Association of America, see the drawings prepared by Henry Wright for the New York State Commission of Housing and Regional Planning, especially the Regional Transect with Sectors & Communities, found at http://www.transect.org/regional_img.html.

[10] Wright's Broadacre City concept dates to the mid-1920s. It was fleshed out in *When Democracy Builds* (Chicago: University of Chicago Press, 1945) and *The Living City* (New York: Horizon Press, 1958).

[11] Le Corbusier, *The City of To-Morrow and Its Planning* (New York: Dover Publications, 1987).

[12] See for instance Jane Fearnley-Whittingstall, *The Ministry of Food* (London: Hodder & Stoughton, 2010).

[13] See http://www.transect.org/rural_img.html.

[14] *Food security* is defined by the United Nations Food and Agriculture Organization as "a condition in which all people, at all times, have physical and economic access to sufficient, safe and nutritious food to meet their dietary needs and food preferences for an active and healthy life." See http://www.ryerson.ca/foodsecurity/definition/fsdefinedmore/index.html for further discussions surrounding the definition of food security.

[15] United States Department of Agriculture, *Household Food Security in the United States, 2009*, Economic Research Report No. ERR-108, by Margaret Andrews, Steven Carlson, Alisha Coleman-Jensen, and Mark Nord (Washington, D.C.: November, 2010). Posted at http://www.ers.usda.gov/Publications/ERR108.

[16] This breaks down into production and initial processing (34 percent); manufacturing, distribution, retail, and cooking (26 percent); and agriculturally induced land use change (40 percent). Eric Audsley, Matthew Brander, Julia Chatterton, Donal Murphy-Bokern, Catriona Webster, and Adrian Williams, *How Low Can We Go: An Assessment of Greenhouse Gas Emissions from the UK Food System and the Scope for Reduction by 2050* (WWF-UK and Food Climate Research Network: United Kingdom, January 2010). Posted at: http://www.fcrn.org.uk/fcrnPublications/publications/PDFs/howlow/WWF_How_Low_Report.pdf.

[17] Ibid.

[18] The term *civic agriculture* was championed by Thomas A. Lyson in *Civic Agriculture: Reconnecting Farm, Food, and Community* (Medford, Mass.: Tufts University Press, 2004).

[19] Janine de la Salle and Mark Holland's *Agricultural Urbanism: Handbook for Building Sustainable Food Systems in 21st Century Cities* (Winnipeg, MB: Green Frigate Books, 2010) offers useful ways to achieve such aims..

[20] Domenic Vitiello and Michael Nairn, "Everyday Urban Agriculture: From Community Gardening to Community Food Security," *Harvard Design Magazine* 30, vol. 2 (fall/winter 2009).

[21] From the public programs statement for the *Living Concrete/Carrot City* exhibit, mounted during fall 2010 at the Sheila C. Johnson Design Center, Parsons The New School for Design, New York, New York. See also http://www.newschool.edu/parsons/pastExhibitions1011.aspx?id=55952.

[22] Christine C. Quinn, City Council Speaker, New York City Council, "FoodWorks: A Vision to Improve NYC's Food System," Press Release #120-2009, Nov. 22, 2010, New York, NY. See also Scott M. Stringer, Manhattan Borough President, "FoodNYC – A Blueprint for a Sustainable Food System," Feb. 2010, at http://www.mbpo.org/uploads/policy_reports/mbp/FoodNYC.pdf.

[23] P. De Graaf & J. W. van der Schans, "Integrated

Urban Agriculture in Industrialised Countries—Design Principles for Locally Organised Food Cycles in the Dutch Context," SASBE09 Conference, Delft, The Netherlands, June 2009. Posted at: http://www.sasbe2009.com/papers.html#papers.

24 Domenic Vitiello and Michael Nairn, "Everyday Urban Agriculture: From Community Gardening to Community Food Security," *Harvard Design Magazine* 30, vol. 2 (fall/winter 2009).

25 *Organopónicos* are urban organic gardens in Cuba. They are often made of low, hollow, concrete walls filled with soil and lines of drip irrigation laid on top of the growing media.

26 For information on the Making the Edible Landscape Project, see http://www.mcgill.ca/mchg/pastproject/edible-landscape.

27 Portions of this introduction were previously included in June Komisar, Joe Nasr, and Mark Gorgolewski, "Designing for Food and Agriculture: Recent Explorations at Ryerson University," *Open House International* 34, No. 2 (June 2009): 61–70.

28 Beginning in fall 2006, a group of fourth-year Ryerson University architecture students, faculty, and outside advisors formed to work on food issues that were embodied in student thesis projects. In this group, researchers shared their strategies and discussed problems and issues regarding food-related architecture and urban design. Concurrently first- and third-year architectural design students were assigned shorter projects that involved local food production through the design of allotment garden plots and greenhouses.

IMAGINING THE PRODUCTIVE CITY

1 Carolyn Steel, *Hungry City: How Food Shapes our Lives* (London: Chatto & Windus, 2008).

2 G. Stanhill, "An Urban Agro-Ecosystem: The Example of Nineteenth-Century Paris," *Agro-Ecosystems* 3 (1977): 269–84.

3 André Viljoen, ed., *Continuous Productive Urban Landscapes (CPULs): Designing Urban Agriculture for Sustainable Cities* (Oxford: Architectural Press, 2005).

4 André Viljoen and Katrin Bohn, "Everything is Continuous: The Continuous Productive Urban Landscape," in *Actions: What You Can Do with the City*, ed. Giovanna Borasi and Mirko Zardini (Montreal: Canadian Centre for Architecture and Amsterdam: SUN, 2008), 205.

5 André Viljoen, Katrin Bohn, and Jorge Peña Diaz, *London Thames Gateway: Proposal for Implementing CPULs in London Riverside and the Lower Lea Valley* (Brighton: University of Brighton and CUJAE Technical University Havana, 2004), 3.

6 Ibid., 22–23.

7 "FarmShare," entry by BK farmyards in the 2009 Buckminster Fuller Challenge.

8 James Howard Kunstler, *The Geography of Nowhere: The Rise and Decline of America's Man-Made Landscape* (New York: Simon & Schuster, 1993).

9 Alan Berger, *Drosscape: Wasting Land in Urban America* (New York: Princeton Arch. Press, 2006).

10 Ann O'M. Bowman and Michael A. Pagano, *Vacant Land in Cities: An Urban Resource* (Washington D.C.: Brookings Institution Center on Urban & Metropolitan Policy, Dec. 2000).

11 See http://www.what-if.info and http://www.geml.info.

12 For a systematic review of the potential land and water surfaces across cities that could serve as resources for urban agriculture, see Chapter 4 of Jac Smit, Joe Nasr, and Annu Ratta, "Urban Agriculture: Food, Jobs and Sustainable Cities" (sponsored by the United Nations Development Programme, unpublished 2nd edition, 2001). Posted at http://www.jacsmit.com.

13 The mission of the Shrinking Cities International Research Network is "to advance international understanding and promote scholarship about population decrease in urban regions and urban decline, causes, manifestations, spatial variations, and effectiveness of policies and planning interventions to stave off decline." See http://www.shrinkingcities.org.

14 See http://www.hantzfarmsdetroit.com.

15 See http://www.recoverypark.org.

16 See David Whitford, "Can Farming Save Detroit?" *Fortune*, December 29, 2009, posted at http://money.cnn.com/2009/12/29/news/economy/farming_detroit.fortune/index.htm; and Catherine Porter, "From Motown to Hoetown," Toronto Star, September 26 2009, posted at http://www.thestar.com/news/insight/article/700654.

17 See http://www.ryerson.ca/foodsecurity/projects/urbandesign/Joongsum%20Kim.pdf.

18 "Park N Farm—Terraforming the Strip Mall Parking Lot," entry by BK farmyards to the 2009 Buckminster Fuller Challenge, posted at www.bkfarmyards.com/philosophy/parknfarm/parknfarm.html.

19 See "Cliffside Plaza: The Post Carbon Strip Mall", entry by Ryerson University students Dov Feinmesser, Aaron Hendershott, Yekaterina Mityuryayeva, and Tommy Tso (advisor June Komisar), Cities Alive 2009 competition. Posted at http://www.greenroofs.org/resources/education_challenge_files/Cliffside_Village.pdf.

20 See http://www.ecocitylab.org/ecocitydesignstudio/ecocities2050/eco-cities/index.html.

21 E.R.A. Architects and the University of Toronto, *Mayor's Tower Renewal—Opportunities Book* (Toronto: City of Toronto, 2008), 82. This initiative began as a graduate thesis by architect Graeme Stewart and was later adopted by the mayor.

22 See http://www.uwsp.edu/cnr/landcenter/tracker/Summer2002/conssubdiv.html.

23 Makoto Yokohari, Marco Amati, Jay Bolthouse, and Hideharu Kurita, "Restoring Agricultural Landscapes in Shrinking Cities: Re-Inventing Traditional Concepts in Japanese Planning," in J. Primdahl and S. Swaffield, eds., *Globalisation and Agricultural Landscapes* (Cambridge: Cambridge UP, 2010), 275.

24 See http://www.natlands.org/services/for-municipalities/case-studies.

25 See http://www.agritopia.com.

26 See http://www.agriburbia.com.

27 See http://www.thetsrgroup.com.

28 Matthew Q. (Quint) Redmond, "Feeding the Future: A New View of 'Providing Lands,'" in *Building Metropolitan Atlanta: Past, Present & Future*, Eighteenth Congress for the New Urbanism (May 2010), 74–77. See also http:// http://www.cnu.org/BuildingAtlanta.

29 Alec Applebaum, "Organic Farms as Subdivision Amenities," *New York Times*, 30 June 2009.

30 Treasure Island is a collaboration between BCV Architects, Conger Moss Guillard Landscape Architects, Hornberger + Worstell, Perkins+Will, and Skidmore, Owings & Merrill.

31 See Smit, Nasr, and Ratta, Chapter 1.

32 Dickson Despommier, *The Vertical Farm: Feeding the World in the 21st Century* (New York: Thomas Dunne Books, 2010). See also www.verticalfarms.com.

33 See this *National Geographic* blog posting for other examples: http://news.nationalgeographic.com/news/2009/06/photogalleries/vertical-farm-towers/photo4.html.

34 See http://www.plantagon.com and http://www.cityfarmer.info/2009/12/06/plantagon-greenhouse-urban-crops-in-a-gigantic-glass-sphere/#more-2960.

35 See http://www.cityfarmer.info/2009/05/20/dragonfly-skyscraper-farm-an-urban-agriculture-proposal-for-new-york/#more-1521.

36 Stan Cox and David Van Tassel, "Why Planting Farms in Skyscrapers Won't Solve Our Food Problems," *AlterNet*, May 3, 2009. Posted at: http://www.alternet.org/module/printversion/146686.

37 ARUP, Sauerbruch Hutton, Experientia, and Galley Eco Capital, "C_Life—City as a Living Factory of Ecology," proposal entered in the Low2No 2009 design competition. See also http://low2no.org/.

38 Raquel Maria Dillon, "Urban Farmers Are Challenging City Halls to Rewrite Ordinances," Associated Press, 5 February 2010, posted at: http://www.cityfarmer.info/2010/02/05/urban-farmers-are-challenging-city-halls-to-rewrite-ordinances/#more-3781.

39 American Planning Association, *Zoning Practice 3: Practice Urban Agriculture*, March 2010.

40 Nancy Kaffer, "Detroit Officials Work to Create Zoning Code for Urban Farming," *Crain's Detroit Business*, 23 March 2010. Also posted at http://www.cityfarmer.info/2010/03/23/detroit-officials-work-to-create-zoning-code-for-urban-farming/#more-4421.

41 See http://www.dott07.com/go/what-is-dott.

42 John Thackara, *Wouldn't It Be Great If…* (London: Dott 07 Design Council, 2007), 82.

43 The Design Council, OneNorth East, Middlesbrough Council, Middlesbrough PCT, Middlesbrough Sure Start, Groundwork South Tees, Middlesbrough Environment City, and the Soil Association assisted the project in a variety of ways. The Minister for the Environment, Food and Rural Affairs subsequently endorsed the project.

44 The Program Director, John Thackara, was sustainability strategist and instigator, David Barrie was in charge of production and regeneration, Debra Solomon, artist, educator, and author of Culiblog (http://www.culiblog.org) provided artistic input to the project, Nina Belk of Zest Innovation provided communication skills, and Architects Katrin Bohn and André Viljoen's *Continuous Productive Urban Landscapes* concept provided an initial theoretical framework.

45 See André Viljoen, ed., *Continuous Productive Urban Landscapes (CPULs): Designing Urban Agriculture for Sustainable Cities* (Oxford: Architectural Press, 2005),1–32.

46 Catherine Early, "Urban Jungle" *The Guardian*, 26 March 2008, 8.

47 Ibid.

48 See http://www.ruaf.org.

49 The ordinance allows for establishment of community gardens on public and privately owned land. Marielle Dubbeling, Elio Di Bernardo, Laura Bracalenti, Laura Lagorio, Virginia Lamas, Marina Rodriguez, Raul Terrile, Antonio Lattura, "Optimisation of the Use of Vacant Land in Rosario," *Urban Agriculture* (Dec. 2003): 24.

50 See http://www.zonneterp.nl/english/index_uk.html.

51 See Alexander van de Beek and Huub ter Haar, *Building with Green and Light* (2009). See also www.kiplant.com/BookPreview.aspx?id=16.

52 The complete policy statement is available online at http://vancouver.ca/commsvcs/southeast/policystatement/index.htm.

53 More about this development available at http://www.thechallengeseries.ca.

54 Holland Barrs, *Designing Urban Agriculture Opportunities for Southeast False Creek* (2007). A link to the firm's final report is available at http://www.hblanarc.ca/projects/project_details.asp?ProjectID=93.

55 The competition-winning team was comprised of Bruce Mau Design Inc.; Inside Outside and OMA (Rem Koolhaas, principal architect), with Oleson Worland Architects.

56 The development maintains a website at http://www.prairiecrossing.com/pc/site/index.html.

57 J.S. Watson, "Preservation of the Environment and Open Space Through Free Market Housing Incentives" (Ph.D. diss., University of Illinois at Chicago, 2006).

58 Nevin Cohen, "The Suburban Farm: An Innovative Model for Civic Agriculture," *Urban Agriculture Magazine* 19 (2007).

59 For a complete description of Innovation Network's activities, see www.innovatienetwerk.org/en.

60 See http://www.mvrdv.nl/#/projects/181pigcity.

61 Experimental Design by Nantapon Junngurn, presented at Silpakorn University in 2009. See http://www.slideshare.net/ru2dstudio/240496-experimental-design-17jul09-slides.

BUILDING COMMUNITY AND KNOWLEDGE

1 Sarah Wakefield, Fiona Yeudall, Carolin Taron, Jennifer Reynolds, and Ana Skinner, "Growing Urban Health: Community Gardening in South-East Toronto," *Health Promotion International* 22, no. 2 (2007): 97.

2 Carolyn Steel, *Hungry City: How Food Shapes our Lives* (London: Chatto & Windus, 2008).

3 Wakefield, Yeudall, Taron, Reynolds and Skinner: 92–101.

4 The University of California Cooperative Extension's guide to starting a community garden can be found at http://ucanr.org/sites/Community_Gardens/files/76430.pdf.

5 Michele Owens, "Gardening to Save Detroit," *O, The Oprah Magazine*, April 2008.

6 See the blog on the Vermont State House garden at http://vtstatehousegarden.wordpress.com.

7 Other examples include the Tate Modern's commission of artist Fritz Haeg to create an Edible Estate on the lawn of a housing project in South London (see page 132 of this volume); the Edible City exhibit at the Netherlands Architecture Institute in 2007, see http://en.nai.nl/exhibitions/archive/2007/item/_rp_kolom2-1_element-tld/1_35402; Stroom den Haag, an art and architecture center in the Hague, which established a two-year, multifaceted program called "Foodprint," see http://www.stroom.nl/activiteiten/manifestatie.php?m_id=4645496; a focus on food within the London Festival of Architecture in 2008; and the Canadian Centre for Architecture's inclusion of gardening as part of its November 2008–April 2009 exhibition *Actions: What You Can Do with the City,* http://cca-actions.org.

8 David W. McMillan, "Sense of Community," *Journal of Community Psychology* 24, no. 4 (1996): 315–25. See also Katherine Alaimo, Thomas M. Reischl, and Julie Ober Allen, "Community Gardening, Neighborhood Meetings, and Social Capital," *Journal of Community Psychology* 38, no. 4 (2010): 497–514.

9 Marcia Caton Campbell and Danielle A. Salus, "Community and Conservation Land Trusts as Unlikely Partners? The Case of Troy Gardens, Madison, Wisconsin," *Land Use Policy* 20, no. 2 (2003): 169–80.

10 UOSF, FTG, Ziegler Design Associates, and UW-Madison Landscape Architecture and Urban and Regional Planning faculty and students facilitated this forum on March 27, 2004.

11 Laura Langston, "From Hockey to Horticulture," *Canadian Gardening* (5 January 2007). Posted at: http://www.canadiangardening.com/gardens/indoor-gardening/from-hockey-to-horticulture/a/1242.

12 See more about the Mars expedition conditions at http://www.asc-csa.gc.ca/eng/astronomy/mars/simulating.asp.

13 The NYRP maintains: "what makes NYRP unique is our unwavering commitment to restore, maintain and introduce innovative youth and adult educational programming and events in our gardens." See www.nyrp.org/Parks_and_Gardens/Community_Gardens for a list of all member gardens. The NYRP is also a steward of parks in New York, encouraging outdoor activities in the parks as well as in community gardens through programs like the NYC Green Stops Partnership that aims to connect children with nature. Activities such as the "Nature in My Neighborhood" youth gardening workshops teach children how to take care of trees, provide basic environmental science knowledge, and promote healthy lifestyles, including outdoor exercise and healthy eating and cooking. In addition, their Teach Green workshops bring NYRP programs to the attention of New York educators, helping them to incorporate the Restoration Project's children's programming into the curriculum.

14 The Trust for Public Land manages additional gardens saved from development. It provides expertise in organizing gardens and community members, and helps to implement fences, water access, and places for the community to gather. Another notable success is the Central Bainbridge Block Association Community Garden in Bedford-Stuyvesant, Brooklyn, a former empty lot now filled with raised beds and amenities including a shaded seating area and a gazebo. See http://www.neighborhoodlink.com/org/bqlt/clubextra/151709586.html for this garden's blog.

15 Anne Raver, "Healthy Spaces, for People and the Earth," *New York Times*, 5 November 2008.

16 As of publication, the organizations with offices at Artscape Wychwood Barns include ANDPVA (The Association for Native Development in the Performing and Visual Arts, b current, Hélène Comay Nursery School, Latino Canadian Cultural Association, Latin American-Canadian Arts Projects, LEAF (Local Enhancement and Appreciation of Forests), New Adventures in Sound Art, The Stop Community Food Centre, Storytelling Toronto, Theatre Direct Canada, and The Wychwood Barns Community Association.

17 Heritage associations involved included Taddlewood Heritage Association, Friends of a New Park, Wychwood Heights BIA, and Hillcrest BIA.

18 Artscape press release, Wednesday, June 20, 2007.

19 "Arts and culture plays a crucial role in building strong and prosperous communities," said The Honourable Caroline Di Cocco, Ontario Minister of Culture, when announcing its $3 million investment in the project. "For artists, finding housing as well as studio space that is affordable and appropriate to their craft in a booming city can be difficult." Artscape press release, Wednesday, June 20, 2007.

20 See http://www.thestop.org.

21 For a full description of the amenities available at The Stop, see http://www.thestop.org/bake-ovens-markets.

22 Metcalf Foundation, *Food Connects us All: Sustainable Local Food in Southern Ontario* (February 2008). Posted at: http://www.metcalffoundation.com/downloads/Food%20Connects%20Us%20All.pdf.

23 The site is owned by the Toronto and Region Conservation Authority (TRCA).

24 Diamond+Schmidt Architects, "Centre for Urban Sustainability: Evergreen Brick Works," pamphlet introducing the project available online at http://www.evergreen.ca/groundbreaking/pdf/EBWCtrForUrbanSustainability.pdf.

25 LEED awards recognize achievement of specific sustainability goals. LEED stands for Leadership in Energy and Environmental Design. A listing of the LEED rating systems is available at: http://www.usgbc.org/DisplayPage.aspx?CMSPageID=222.

26 See http://www.mcgill.ca/mchg/pastproject/edible-landscape, http://www.santropolroulant.org, and http://www.alternatives.ca.

27 V. Bhatt, L. M. Farah, N. Luka, and J. M. Wolfe, "Making the Edible Campus: A Model for Food-Secure Urban Revitalization," *Open House International* (2008): 82.

28 Ibid.

29 V. Bhatt, L. M. Farah, N. Luka, J. M. Wolfe, R. Ayalon, I. Hautecoeur, J. Rabinowicz, and J. Lebedeva, "Reinstating the Roles and Places of Productive Growing in Cities," *The Sustainable City V: Urban Regeneration and Sustainability* (Southampton, U. K.: Transactions of the Wessex Institute, 2008).

30 Bhatt, Farah, Luka, and Wolfe, "Making the Edible Campus," 82.

31 Ibid.

[32] From the 2008 National Urban Design Award listing. See also http://www.mcgill.ca/mchg.

[33] Chuck Sudo, "Growing a Home: Urban garden thrives during its first winter in Englewood," *Chicago Sun-Times*, 19 March 2008. Posted at http://www.highbeam.com/doc/1N1-11F8328A07C9CA28.html.

[34] Orrin Williams, "Growing Home and the Emergence of Urban Agriculture in Chicago," *Urban Agriculture Magazine* 18 (2007): 36-37. Posted at: http://www.ruaf.org/sites/default/files/Article%2012.pdf.

[35] Piers Taylor, personal correspondence.

[36] From the project description by Amale Andraos and Dan Wood, WORK Architecture Company. See also http://work.ac/pf-1.

REDESIGNING THE HOME

[1] United Nations, Department of Economic and Social Affairs, "World Urbanization Prospects: The 2005 Revision." Working paper, 2005. Complete download available at: http://www.un.org/esa/population/publications/WUP2005/2005WUPHighlights_Final_Report.pdf.

[2] Caesar B. Cororaton and Erwin L. Corong, "Philippine Agricultural and Food Policies: Implications for Poverty and Income Distribution," Research Report 161 (International Food Policy Research Institute: Washington, D.C., 2009).

[3] "Heaven on Earth: The Plan of St. Gall," *The Wilson Quarterly* 4, no. 1 (Winter 1980): 171–79.

[4] Witold Rybczynski, *Home: A Short History of an Idea* (New York: Viking, 1986), 60.

[5] Food and Agriculture Organization of the United Nations, "Box gardens for vulnerable households, Kitwe, Zambia." See ftp://ftp.fao.org/SD/SDA/SDAR/sard/English%20GP/EN%20GP%20Africa/Box_gardens_%20Zambia.pdf.

[6] Alex Frangos, "The Green House of the Future," *Wall Street Journal*, 27 April 2009. Posted at: http://online.wsj.com/article/SB124050414436548553.html.

[7] See http://www.verticalfarm.com/designs.

[8] See http://www.mole-hill.ca/?page_id=47 for more on the history of this project.

[9] Sean McEwen, personal correspondence, 25 August 2010.

[10] Sean McEwen leads an annual tour for the Urban Agriculture class at the University of British Columbia's (UBC) School of Community and Regional Planning, as well as tours for a UBC Community Development class and for Heritage Vancouver.

[11] Thor Lewis modeled the shed and John Freeman of ERTH Design Consultants, along with a crew of volunteers and staff from Mountain Equipment Co-op, designed and built it.

[12] *Urban Agriculture Notes*, http://www.cityfarmer.org, which has been published continuously since 1994, has been succeeded by City Farmer News, http://www.cityfarmer.info.

[13] See http://www.formshiftvancouver.com for further information on this competition.

[14] See http://www.architecture2030.org for more on this organization's ongoing initiative to achieve a dramatic reduction in greenhouse gas emissions by the building sector.

[15] *Harvest Green Project by Romses Architects*. Reported on ArchiCentral.com, 12 May 2009.

[16] Kongshaug served previously as a project manager for Making the Edible Landscape, a four-year program leveraging urban agriculture as a means of upgrading the quality of life for citizens of four cities in dense urban locations (see page 38). He was also co-editor of *EL1: Making the Edible Landscape: A Study of Urban Agriculture in Montreal* (Montreal: McGill University, 2005).

[17] Michael Pollan, *Second Nature: A Gardener's Education* (New York: Grove, 1991), 65–78. Reprinted in Fritz Haeg, *Edible Estates: Attack on the Front Lawn* (New York: Metropolis, 2008), 34.

[18] Curated by Stacy Switzer.

[19] The Land Institute is a research, education and policy organization dedicated to sustainable agriculture.

[20] Produced by the New York Restoration Project in partnership with Friends of the High Line for Hudson Guild, a community center, Robert Fulton Houses and Elliott-Chelsea House, New York, New York.

[21] For more on this competition, see http://www.livingsteel.org.

[22] See http://cascadiagbc.org.

[23] The Living Building Challenge is an effort by the Cascadia Green Building Council to develop a standard for buildings with negligible impact on the environment.

[24] See http://ilbi.org for a definition of individual prerequisites.

[25] Zach Mortice, "From Farm to Market, Down the Stairs, Around the Block," *AIArchitect*, 26 October 2007.

PRODUCING ON THE ROOF

[1] Stephen W. Peck et al., *Greenbacks From Green Roofs*, a special report prepared for the Canada Mortgage and Housing Corporation (Canada: CMHC, 2001). A download of this document, which compares the advantages and disadvantages of extensive and intensive green roofs, is available at http://www.cmhc-schl.gc.ca/odpub/pdf/62665.pdf?fr=1303397364855.

[2] Daniel Roehr, "Rooftop Agriculture: Greenroofs as Productive Envelopes," working paper, Cities Alive Conference 2009, Toronto, Canada, 2009. A download is available at http://www.scribd.com/doc/34475874/Growing-Fruits-and-Vegetables-on-Rooftops. See also http://www.greenskinslab.sala.ubc.ca for information on the University of British Columbia School of Architecture's ongoing research into green roofs.

[3] See http://www.cityfarmer.info/2008/05/26/rocket-science-%E2%80%93-an-edible-rooftop-garden-in-portland.

[4] S. Alward, R. Alward, and W. Rybczynski, *Rooftop Wastelands*, special report prepared for the Minimum Cost Housing Group at McGill University (Montreal: McGill University, 1976). Posted at: http://www.mcgill.ca/mchg/pastproject/rooftop.

[5] See www.mcgill.ca/mchg/pastproject/rooftop.

[6] Martin Price, "Rooftop Gardening in St. Petersburg," *Urban Agriculture Magazine* 10 (August 2003): 17.

[7] See http://gothamgreens.com for the organization's current initiatives and events.

[8] Alison Macgregor, "Commercial Rooftop Garden in Montreal is World's First," *Montreal Gazette*, 28 October 2010. Posted at: http://www.cityfarmer.info/2010/10/29/commercial-rooftop-garden-in-montreal-is-worlds-first.

[9] Alyssa Danigelis, "Harvest Produce at the Grocery Store," *Discovery News*, 1 September 2010. Posted at http://news.discovery.com/tech/harvest-produce-at-the-grocery-store.html.

[10] Valerie Easton, " At Seattle's Bastille, the Garden Goodies Are on the Roof," *Seattle Times*, 15 November 2009. Posted at: http://seattletimes.nwsource.com/html/pacificnw/2010203442_pacificplife15.html.

[11] Annie Novak also runs Growing Chefs, a food-education, nonprofit organization that connects people to food "from field to fork."

[12] Megan Paska of Brooklyn Honey and Annie Novak keep the bees.

[13] The team consisted of Gwen Schantz, a cofounder of Bushwick Food Cooperative and farm manager at Roberta's, Chris Parachini and Brandon Hoy, co-owners of Roberta's, and Anastasia Plakias.

[14] Green design expert Monica Kuhn led this charrette. See her firm's website at http://www.mekarch.ca.

[15] Groups involved included Carrot Cache, along with the participation of a large number of community associates and funding partners, as well as incentives from Toronto's Eco-Roof Incentive Program.

[16] Research will be led by Dr. Youbin Zheng, an adjunct professor in the School of Environmental Sciences at the University of Guelph. This project will train several graduate students, who will be integrated into the research team of the Controlled Environment Systems Research Facility (CESRF) at that institution.

[17] Conestoga-Rovers & Associates, Mariko Uda, and CSR+ Vermicast undertook these studies, respectively.

[18] Professional development and design services are provided by Jonathan Rose Companies, which became the project manager in 2006, cooperating with Murphy, Burnham & Buttrick Architects and Robert Silman Associates Structural Engineers. These groups coordinate with P.S. 41's administration, faculty, and parent-teacher association.

[19] The offices of Manhattan Borough President Scott Stringer, City Council Speaker Christine C. Quinn, and New York State Senator Thomas K. Duane, the P.S. 41 parent-teacher association, the Environmental Resource Management (ERM) Foundation, and others have provided funding.

[20] The office of Manhattan Borough President Scott Stringer is contributing $500,000 toward the cost of the $750,000 project.

[21] A public-private partnership between the school, parents, and outside funding financed the transformation of the rooftop.

[22] The architects Edelman Sultan Knox Wood designed the building, located at 1323 Louis Nine Boulevard in the Bronx. The Women's Housing and Economic Development Corporation (WHEDCo) developed the overall site, along with neighboring Intervale Green, a 127-unit affordable housing development.

[23] The Parsons Design Workshop occupies a key position in the curriculum of the Parsons Master

of Architecture program and introduces real-world constraints into comprehensive design. The Workshop also incorporates issues of basic design, communal and social spaces, and the "natural" and energetic into its studio sequence, introducing students to regulations, consultants, and clients.

24 Support was provided by architects Randy Wood and Andrew Knox to ensure compliance with local building code and filing protocols for construction issues related to the roof structures, and by Billie Cohen on issues of landscape design.

COMPONENTS FOR GROWING

1 Sevanatha, *The Case of Colombo, Sri Lanka* (Sri Lanka: Urban Resource Centre: 2003). Posted at: http://www.ucl.ac.uk/dpu-projects/Global_Report/pdfs/Columbo.pdf.

2 As part of this program, GrowNYC trains volunteers from the Green Apple Corps and Million Trees NYC Training Program. This is part of a larger strategy for public education about water use, including a manual with instructions for creating individual rainwater diversion systems. See http://www.grownyc.org/openspace/rainwater.

3 A very different approach can be found in Ariana-Soukra, a suburb of Tunis, Tunisia, where part of a school's grounds became a new educational garden. To avoid using clean water for irrigation, the plumbing was modified to capture graywater that is fed into four underground cisterns that supply the garden. This project, funded by the Canadian International Development and Research Centre, is coordinated by Moez Bouraoui. See http://www.agriurbanisme.org.

4 See a further description of how this product works at http://www.leevalley.com/US/garden/page.aspx?p=47098&cat=2,33140&ap=1.

5 Additional examples include the Korean-made Red Dragon and the indoor composter from San Francisco–based NatureMill. The latter can absorb about 120 pounds of food waste—three times its weight—each month, while using less than 5 kilowatt-hours per month. See http://www.cityfarmer.info/2010/05/07/the-electric-red-dragon-a-new-type-of-composter and http://www.naturemill.com.

6 A number of examples can be found at websites such as http://www.unclejimswormfarm.com.

7 See http://www.thewormdude.com for a complete description of this product.

8 Such systems include the Worm Chalet from Cathy's Crawly Composters (Canada) http://www.cathyscomposters.com; Clean Air Gardening's Worm Composter Vermicomposting Bin (Australia), http://www.cleanairgardening.com/worcomverbin.html; and Can-o-worms and Vesey's Worm Factory Composting Bin System (U.S.A.), http://www.veseys.com/ca/en/store/tools/composting/thewormfactory.

9 Rocco Rossi, former Mayoral Candidate quoted during the 2010 Election in the article by Don Peat and Ian Robertson, "Chickens an Election Issue?" *Toronto Sun*, 28 April 2010.

10 See http://ecolocalizer.com/2009/08/08/beehaus-and-eglu-promote-urban-agriculture-in-europe-very-local-food for more on the Eglu.

11 Sponsored by the Holcim International Foundation. Awards are given to competition projects that showcase the importance of sustainability in construction. See http://www.holcimfoundation.org/T989/Regional_Holcim_Awards_2010.htm for past award winners.

12 Nigel Dunnett and Noël Kingsbury suggest that steep mountainsides can be considered the natural precursors to many contemporary hydroponic systems. Water running down a mountainside absorbs nutrients and passes them on to the roots of mountainous plants below in much the same way as many contemporary living-wall systems work today. Nigel Dunnett and Noël Kingsbury, *Planting Green Roofs and Living Walls*, 2nd ed. (Portland, Ore.: Timber Press, 2003).

13 See http://www.brightfarmsystems.com.

14 See http://gothamgreens.com.

15 See http://www.allaboutfeed.net/photos/farm-visit-happy-shrimp-farm-in-the-netherlands-3073.html.

16 See http://sweetwater-organic.com/blog.

17 See http://www.windowfarms.org.

18 Wayne Roberts, "Northwest Territories Gardeners and Farmers Work Together for Local Food," *NOW Magazine*, 26 August 2010.

19 This is one of sixteen new gardens created by an initiative called Grow It, Eat It, Move It, Live It (GEML) in the span of a year and a half. GEML is a municipal initiative that is creating new sites of production on city-owned land in a district of Birmingham, England, the Ladywood Constituency. See http://www.geml.info.

20 See http://www.myownbackyard.ca.

21 See a video of the installation at http://www.youtube.com/watch?v=dJypFEA33lU.

22 See http://www.what-if.info.

23 See http://www.cityfarmer.info/2010/03/27/kenya-bag-an-urban-farm/#more-4495 for more about this project in Kenya.

24 See http://www.what-if.info.

25 See http://www.tdelaney.com/Commissions/public.html.

26 See also http://lisboasos.blogspot.com/2009/04/united-colors-of-leonel-moura.html.

27 The Biotop system, developed in Montreal, adapts the container principle for roofs. It consists of a slotted plant container filled with vermiculite and potting soil nested inside a 10-liter water-retention container made from recyclable plastic resin. The vermiculite layer decreases spiral root formations, which occur regularly in non-copper-coated traditional pot cultures. This patented system promotes the development and differentiation of dense, ramified root systems by providing continuous inputs of mineral nutrients via the soil, and perpetual access to water from below. When full, the container adds 9 pounds per square foot of weight to the roof structure, far less than typical green roof membrane systems. Joined containers create a common water reserve and a water-irrigation system that can be fully automated. See http://www.biotopcanada.com.

28 The manual is available in English and French at http://rooftopgardens.ca/files/manual%202009%20web.pdf.

29 EarthBox has been based in Scranton, Penn., since 1994. See http://www.earthbox.com.

30 See http://www.thegrowingconnection.org.

31 Another type of hydroponic vertical growing system, the Tower Garden by Future Growing, was designed for tall, continuous heights. Even though it is made up of pieces that are assembled, this is not truly a stacked system—rather, it is a perforated long tube. Plants are placed into slots in the tube, which sits on a 25-gallon reservoir. An internal system recirculates water and dispenses nutrients, running for three minutes out of every twelve. The Tower Garden is made from food-grade plastic that has been UV stabilized for years of outdoor use. This has enabled its use, starting in the summer of 2010, on the rooftop of an apartment building in New York's Greenwich Village by the owner/chef of a new restaurant on the building's ground level. The system of 60 short towers contains over 1,000 plants, and uses a small solar panel to provide power. The first harvest took place only four weeks after the system was installed. See http://mytowergarden.com. Future Growing has other types of very tall towers that are intended for commercial production; see http://www.futuregrowing.com.

32 Herbs, a high-value crop heavily used in the store's prepared foods section, were chosen as ideal produce to grow in a space with limited light conditions. See http://brightfarmsystems.com/completed-and-built/whole-foods-market-usa.

33 Hydrogardenshop has created a similar system. See http://www.ezgro.com and http://www.hydrogardenshop.com.

34 Mark Bearak, Dora Kelle, and Adam Mercier are the designers. See their blog at http://amphorae.wordpress.com.

35 See http://www.what-if.info.

36 See http://agreenroof.com/urban-farms/mobile-edible-walls.

37 See http://www.barthelmes.com.

38 Ken Belson, "The Rooftop Garden Climbs Down a Wall," *New York Times*, 18 November 2009. Posted at: http://www.nytimes.com/2009/11/19/business/energy-environment/19WALLS.html?_r=3&ref=organic_gardening.

39 ELT Easy Green, http://www.eltlivingwalls.com.

40 See http://www.bacsac.fr/en.

41 See http://cart.woollypocket.com/Wally-One.

ACKNOWLEDGMENTS

We are very grateful to the research assistants who helped us directly in the preparation of this book by drafting cases, obtaining images, and seeking permissions. These include the following: Sebastian Lubczynski, Carolin Mees, Nicholas Potovszky, Nadia Qadir, and Abra Snider.

In addition to this core team, a much larger number of people enabled the realization of this book by providing hundreds of images and permissions for their use—this book could not have existed without them. We are also grateful for the persistence, the sharp eye, and the tireless effort of our editor at The Monacelli Press, Stacee Gravelle Lawrence, her colleague Rebecca McNamara, who proved invaluable in ironing out so many details, and our book designer, John Clifford of Think Studio.

This is a book of case studies involving dozens of sources. It could not have come about without the generous help of the creators of these projects and other stakeholders. The work on the book builds on the broader Carrot City initiative, which has been based at Ryerson University over the past few years. The funding and in-kind contributions for this initiative, particularly from Ryerson University and its CFI-funded REAL lab, were essential to making it possible. The initiative started modestly when we advised architecture students who integrated urban agriculture into their undergraduate theses in 2006–8. These students laid the groundwork for this project by demonstrating the need to study the ways that design is connected to enabling the production of food, particularly in urban locations.

This led to a symposium on the role of food and agriculture in the design and planning of buildings and cities, which took place in May 2008. The success of this symposium led to the exhibition on which this book is directly based, held at the Design Exchange in Toronto in the winter of 2009. Variations of this exhibition have subsequently been shown in various locations internationally. We wish to thank our dedicated team of graduate and undergraduate students who worked countless hours to make that original exhibit happen. Their research for the exhibit boards formed the basis for many of the case studies that are featured in this book. They include: Krysia Gorgolewska, Antonio Leung, Jun Liu, Pamela Love, Stanley Wai Lung, Danielle O'Donoghue, Rachel Pressick, Liming Qiu, Irivia Rovika, Julie Jooyun Shin, Adam Smith, Melody Taghi-Poor, Pearl Waiyin Tam, Micah Vernon, and Elmira Yousefi.

Beyond this team who worked on the initial case studies, an extensive number of individuals and organizations supported the development of the exhibit, the inclusion of objects in it, its transfer to other venues, and a number of other roles as part of this initiative. The acknowledgment for these contributions can be found at the initiative's website: www.carrotcity.org.

One website in particular proved to be invaluable as a resource for identifying case studies and keeping up with the latest developments in design for urban agriculture: www.cityfarmer.info. We are grateful to its creator, Mike Levenston, for maintaining such a knowledge treasure. Two of the authors also thank Nevin Cohen at The New School for hosting us in New York City during a critical period of the writing of the book in winter 2010.

Finally, we would like to thank our family and friends for their patience as we worked on completing this project.

PROJECT CREDITS

Middlesbrough Urban Farming Project
Client: DOTT 07 (Design Council/OneNorth East) and the Middlesbrough Borough Council
Designers: John Thackara, Program Director, Dott 07; David Barrie, David Barrie & Associates; Nina Belk, Zest Innovation; Debra Solomon, artist and Culiblog author; Bohn & Viljoen Architects; the citizens of Middlesbrough
Related Links: http://www.borohealthytown.com; http://www.dott07.com

Ravine City
Designers: Chris Hardwicke and Hai Ho, Sweeny Sterling Finlayson & Co Architects
Related Links: http://www.andco.com/

Gardiner Urban Agriculture Hub
Designer: Andy Guiry, Ryerson University thesis

Making the Edible Landscape
Designers: Edible Landscape Project of the Minimum Cost Housing Group, McGill University; International Network of Resource Centres on Urban Agriculture & Food Security (RUAF); ETC-Urban Agriculture, The Netherlands
Related Links: http://www.mcgill.ca/mchg/pastproject/edible-landscape; http://www.ruaf.org; http://www.etc-urbanagriculture.org/

Greenhouse Village
Designers: Innovation Network with Dr. Noor van Andel, Prof. J. Kristinsson and Prof. G. Lettinga
Related Links: http://www.zonneterp.nl; http://www.innovatienetwerk.org

Southeast False Creek
Client: City of Vancouver
Designers: PWL Partnership Landscape Architects, Inc. (overall landscape design); Durante Kreuk Ltd. (Millennium Water landscape design)
Additional Participants: HB Lanarc
Related Links: http://vancouver.ca/commsvcs/southeast/

Parc Downsview Park
Client: Parc Downsview Park, Inc.
Designers: Tree City Concept Team: Mau Design Inc., Inside Outside, OMA, and Oleson Worland Architects; Jane Hutton (FoodCycles site on Cultivation Campus); Blair Robins and Ian Lazarus (Hedgerow Project); Megan Albinger (747 Community Greenhouse Proposal)
Additional Participants: FoodCycles
Related Links: www.pdp.ca

Prairie Crossing
Client: Prairie Holdings Corporation; George and Vicky Ranney, Co-developers
Related Links: http://www.prairiecrossing.com

Agroparks
Designer: Innovation Networks
Related Links: http://www.innovatienetwerk.org/en/concepten/view/63/Agricentre.html

Pig City
Client: Stroom, The Hague's Center for Visual Arts
Designer: MVRDV
Additional Participants: Jan-Willem van der Schans, Agriculture Economics Research Institute LEI, Wageningen NL
Related Links: http://www.mvrdv.nl

Troy Gardens
Client: Community GroundWorks at Troy Gardens
Designers: Ziegler Design Associates; JJR
Additional Participants: Madison Area Community Land Trust; Urban Open Space Foundation/Center for Resilient Cities; University of Wisconsin–Madison Department of Landscape Architecture
Related Links: http://www.troygardens.org

Inuvik Community Greenhouse
Client: Community Garden Society of Inuvik
Related Links: http://www.inuvikgreenhouse.com

Curtis "50 Cent" Jackson Community Garden
Client: New York Restoration Project
Designer: Walter J. Hood
Additional Participants: G-Unity Foundation
Related Links: http://www.nyrp.org/Parks_and_Gardens/Community_Gardens/Queens/Curtis_50_Cent_Jackson; http://www.wjhooddesign.com

Artscape Wychwood Barns
Client: Artscape and The Stop Community Food Centre
Designer: Joe Lobko, du Toit Allsopp Hillier | du Toit Architects Ltd.
Related Links: http://www.torontoartscape.on.ca/places-spaces/artscape-wychwood-barns; http://www.thestop.org

Evergreen Brick Works
Client: Evergreen
Architecture: Joe Lobko, du Toit Allsopp Hillier | du Toit Architects Ltd.; Michael Leckman, Diamond + Schmitt Architects; Edwin Rowse, E.R.A. Architects
Landscape Architecture: John Hillier, du Toit Allsopp Hillier; Claude Cormier, Claude Cormier Architectes Paysagistes, Inc.
Signage and Graphic Design: Debbie Adams, Adams + Associates, Inc.
Additional Participants: Toronto and Region Conservation Authority
Related Links: http://ebw.evergreen.ca

Niagara Community Food Centre
Designer: Jordan Kemp Edmonds, Ryerson University thesis

The Science Barge
Designer: New York Sun Works
Additional Participants: Groundwork Hudson Valley
Related Links: http://nysunworks.org/thescience-barge; http://www.groundworkhv.org/programs/environmental-education/science-barge

The Edible Schoolyard
Client: Edible Schoolyard New York at the Arturo Toscanini School, P.S. 216
Designer: WORK Architecture Company
Related Links: http://work.ac/ps-216-edible-schoolyard

The Edible Campus
Client: McGill University
Designer: The Minimum Cost Housing Group at McGill University
Additional Participants: Santropol Roulant; Alternatives
Related Links: http://www.mcgill.ca/mchg/; http://www.santropolroulant.org/2009/F-home.htm; http://www.alternatives.ca/en

Wood Street Urban Farm
Client: Growing Home
Designer: SHED Studio
Related Links: http://www.growinghomeinc.org;
  http://www.shedchicago.com

Leadenhall Street City Farm
Client: British Land
Designer: Mitchell Taylor Workshop
Related Links: http://www.mitchelltaylorworkshop.
  co.uk

Public Farm One
Client: Museum of Modern Art
Designer: WORK Architecture Company
Additional Participants: LERA Structural Engineering;
  Art Domantay Construction; Queens County Farm
  Museum; GrowNYC; Horticultural Society of New York
Related Links: http://www.publicfarm1.org

REDESIGNING THE HOME

Mole Hill Community Housing
Clients: City of Vancouver; BC Housing; Mole Hill
  Community Housing Society
Designers: DIALOG; Sean R. McEwen Associated
  Architects; Durante Kreuk Lanscape Architect
Additional Participants: Mole Hill Living Heritage
  Society; The Friends of Mole Hill
Related Links: http://www.mole-hill.ca

City Farmer
Client: City Farmer
Designer: Michael Levenston
Related Links: http://www.cityfarmer.info; http://
  www.cityfarmer.org

Harvest Green
Designer: Romses Architects
Related Links: http://www.romsesarchitects.com

Maison Productive House
Designer: Produktif Studio de Design
Additional Participants: Design 1 Habitat
Related Links: http://productivehouse.com; http://
  produktif.com; http://www.design1habitat.org

Edible Estates Regional Prototype Gardens
Designer: Fritz Haeg
Related Links: http://www.fritzhaeg.com/garden/
  initiatives/edibleestates/main.html

Edible Terrace
Designers: Anthony Campbell and James West,
  Manchester School of Architecture student project

Agro-Housing
Designer: Knafo Klimor Architects
Related Links: http://www.kkarc.com

Center for Urban Agriculture
Designer: Mithun Architects, Planners and Designers
Related Links: http://mithun.com

60 Richmond Street East Housing Co-operative
Client: Toronto Community Housing Corporation
Designer: Teeple Architects Inc.
Related Links: http://www.torontohousing.ca/
  investing_buildings/regent_park/60_richmond_
  street_east; http://www.teeplearch.com/

PRODUCING ON THE ROOF

New York City Rooftop Farms
Eagle Street Rooftop Farms
Client: Eagle Street Rooftop Farms; Broadway Stages
Designers: Ben Flanner and Annie Novak; Goode
  Green
Related Links: http://www.rooftopfarms.org
Brooklyn Grange
Client: Brooklyn Grange Farm
Designers: Ben Flanner, Brandon Hoy, Chris
  Parachini, Anastasia Plakias, and Gwen Schantz;
  Bromley Caldari Architects
Related Links: http://www.brooklyngrangefarm.com;
  http://bromleycaldari.com/

New York City Rooftop Greenhouses
Forest House
Client: Blue Development Group
Designer: BrightFarm Systems
Related Links: http://brightfarmsystems.
  com/scheduled-for-construction/
  blue-sea-developments-new-york
Eli Zabar's Vinegar Factory
Client/Designer: Eli Zabar
Related Links: http://www.elizabar.com

Uncommon Ground Restaurant
Client: Uncommon Ground
Designer: Michael Cameron
Related Links: http://www.uncommonground.com;
  http://www.eatthisgrowthat.blogspot.com

Carrot Green Roof
Client: Carrot Common
Designer: Tafler Rylett
Additional Participants: Carrot Cache; Carrot Green
  Roof Project Team; Natvik Ecological; University of
  Guelph Landscape Architecture Program
Related Links: http://www.carrotcommon.com

New York Rooftop School Gardens
P.S. 41
Client: The Earth School; Tompkins Square Middle
  School; P.S. 41
Designers: Michael Arad, Handel Architects; Joseph
  Donovan, Stantec Architects
Additional Participants: Vicki Sando, Christopher Hayes
Related Links: http://www.theearthschool.org; http://
  www.handelarchitects.com; http://www.stantec.com
P.S. 64
Client: The Greenwich Village School, P.S. 64
Designers: Jonathan Rose Companies; Murphy
  Burnham & Buttrick Architects; Mark Vetter
Related Links: http://www.rose-network.com; http://
  www.mbbarch.com
P.S. 333
Client: Manhattan School for Children, P.S. 333
Designers: BrightFarm Systems; Kiss + Cathcart,
  Architects
Related Links: http://brightfarmsystems.com/
  completed-and-built/manhattan-school-for-
  children-usa; http://www.kisscathcart.com/

Gary Comer Youth Center
Client: Comer Science & Education Foundation
Designers: Hoerr Schaudt Landscape Architects;
  John Ronan Architect
Related Links: http://www.gcychome.org/architec-
  ture.html; http://www.hoerrschaudt.com; http://
  www.jrarch.com/

BronXscape, Louis Nine House
Client: Neighborhood Coalition for Shelter
Designers: David J. Lewis/Parsons The New School
  for Design
Related Links: http://www.ncsinc.org/

Fairmont Hotel Gardens
Client: Fairmont Hotels and Resorts
Related Links: http://www.fairmont.com/waterfront/
  guestservices/restaurants/herbgardenhoneybees.htm

Brownstone Roof Garden
Designer: Jeff Heehs

IMAGE CREDITS

Numbers refer to page numbers.

Ray Adams 108–109
Megan Albinger 49 bottom
Michael Arad, Handel Architects 176 bottom left, 176 bottom right, 177 top right, 177 bottom
Tom Arban/DTAH 76 bottom
David Barrie 27 bottom
Mark Bearak, Dora Kelle, and Adam Mercier 221 left, 221 right
Vikram Bhatt/MCHG, McGill Univ. 96 top left
Courtesy BioTop 222 top, 222 bottom right
Hedrick Blessing 53 bottom
Mike Blois 18 left, 18 right
Bohn & Viljoen Architects 17, 26, 27 top, 27 bottom, 28 top left, 28–29, 207 top, 207 bottom
Boston Architectural College Urban Design/Build Studio 2007 60 top right
Courtesy Marc Boucher-Colbert 155 right
Bright Farm Systems 164, 165 top, 165 bottom, 179 top, 223
Bright Farm Systems/Kiss+Cathcart 178 top, 178 bottom, 210 left, 210 right, 212 all
Courtesy Bromley Caldari Architects 163 top
Brook McIlroy/Parc Downsview Park, Inc. 47
Michael Cameron/Uncommon Ground Restaurant 168, 169 top
Anthony Campbell and James West 136, 137 top, 137 bottom, 138, 139
Andrew Collings, Growing Home 99 bottom
Claude Cormier Architectes Paysagistes 80 center, 80 bottom, 83 top, 83 center
Kimberly Curry/Carrot Common 170
Courtesy Christopher Delaney 217, 218 top, 218 bottom
Alex De Cordoba 231 top
Diamond+Schmitt 78 left, 82, 83 bottom left, 83 bottom right
Durante Kreuk Ltd. 45
Du Toit Allsopp Hillier | Du Toit Architects 75, 76 top, 78–79, 79 bottom right
Courtesy Edible Vancouver 155 left
Courtesy ELT Easy Green 227, 228 left, 228 right
Jordan Kemp Edmonds 84, 85 top, 85 bottom
Eric Ellingsen, Homero Rios, and Mo Phala 23
Courtesy Evergreen Brick Works/Bernice Gardner 79 bottom left
Courtesy Evergreen Brick Woks/Ferruccio Sardella 80–81
Courtesy Fairmont Hotels 188, 189 all, 190 top left, 191 top left
Leila Marie Farah/MCHG, McGill Univ. 95 top right, 95 bottom, 96 bottom
Dov Feinmesser, Aaron Hendershott, Katrina Mitt, and Tommy Tso 22
Elizabeth Felicella 104, 105 top, 106 left
Leslie Furlong 132 left, 132 right, 133 top
Courtesy Jerry Gach/Worm Inn 202 top
Shai Gil/Teeple Architects 150 left, 150 right

Oto Gillen 135 bottom
Gordon Graff Studio 114
Courtesy Green Living Technologies 229 top, 229 bottom
Raef Grohne/DIALOG 116 left, 116 right, 117 top, 119 bottom
Courtesy GrowNYC 198
Andy Guiry 34, 35 top, 35 bottom, 36 bottom, 36–37
Fritz Haeg 134, 135 top left and right
Chris Hardwicke and Hai Ho 30, 31, 32, 33 top, 33 bottom
Ismael Hautecoeur/Alternatives 96 top right, 224 top right, 224 bottom, 225 bottom right
Alec Hawley/The Stop 77 top, 77 bottom
Jeff Heehs 192 right
Courtesy Henry Holt & Company 111
Courtesy Hoerr Schaudt Landscape Architects 182 top, 182 bottom
Walter J. Hood 2–3, 70, 71 top, 71 bottom left, 71 middle right, 71 bottom right, 72 top, 72 bottom, 73
Gretchen M. Hooker 219 top, 219 bottom left
Jane Hutton 49 top left
Zora Ignjatovic/Carrot Common 171 top, 171 bottom
Innovation Network, The Netherlands Ministry of Agriculture, Nature and Food Quality 41, 42 top, 42 bottom, 43 top, 43 bottom, 54, 55 left, 55 right
Jesse Colin Jackson 196 top left, 196 top center, 196 top right
Courtesy JJR, MACLT 65
Courtesy Just Food 205 top
Courtesy Kiss+Cathcart 179 bottom, 206
Courtesy Knafo Klimor Architects 140, 141 top, 141 bottom, 142 all, 143 all
Lynne Krisfalusi, Parc Downsview Park Inc. 48 left
Andy Kropa 204
Sergey Kuznetsov 112
Courtesy Lee Valley Hardware 200 left
David J. Lewis/The New School 184 right, 185 top, 185 bottom, 186 bottom
Donnelly Marks 161, 162 top
Sean R. McEwen Associated Architects 117 bottom, 118 top, 118 bottom
Minimum Cost Housing Group, McGill University 38, 39, 40 all, 195
Courtesy Mitchell Taylor Workshop 100, 101 all, 102–103
Courtesy Mithun Architects, Planners, and Designers 144, 145 top, 145 bottom, 146 top, 146 bottom, 147
Michael Moran 184 left, 186 top, 187
Philippe Morin 67, 68 top, 68 bottom, 69 top, 69 bottom
Tegan Moss and Tom Hutchinson 157
Stacey Murphy 19
MVRDV Architects and Wieland & Gouwens 56 left, 56 top right, 56 bottom right, 57 top, 57 bottom
Joe Nasr/June Komisar 49 top right, 59 left, 59 right, 60 left, 60 bottom right, 62 left, 62 right, 63, 64 left, 64 right, 86, 87 top, 87 bottom, 88 left, 88 right, 89 top, 94, 95 top left, 97, 99 center right, 119 top, 120 left, 120 right, 122 top, 122 bottom, 166 top, 166 bottom,

167 top, 167 bottom, 173 bottom left, 173 bottom right, 190 top right, 191 top right, 191 bottom left, 191 bottom right, 192 left, 193 left, 193 right, 196 bottom, 201 top right, 201 bottom, 202 bottom, 205 center left, 205 bottom left, 211 bottom, 214 left, 214 right, 216, 222 bottom left, 224 top left, 225 top left, 225 bottom left, 225 top right, 225 center right, 226 top, 226 bottom, 230 right
Courtesy NatureMill, Inc. 200 right, 201 top left
Natvik Ecological/Masha Kazakevich 172 top
Barbara Norman 175 top
Nutri-Centre LaSalle 215 bottom
Courtesy NY Sun Works 89 bottom
Scott Nyerges 158 left, 158 right, 159, 160 top, 160 center, 160 bottom
Okrent Associates 183
Courtesy Omlet Ltd. 203, 205 bottom right
Zoran Orlic 169 center right, 169 bottom left
Phipps Rose Dattner Grimshaw 115
Suthi Picotte 231 bottom
Anastasia Plakias/Brooklyn Grange 162 top left, 162 bottom, 163 bottom
Courtesy Plantagon International AB 25
Prairie Holdings Corporation 51, 53 top
Courtesy Produktif Studio de Design 126, 127 top, 127 bottom, 128 all, 129 top, 129 bottom, 130 top, 130 bottom, 131 top, 131 bottom
Courtesy PWL Partnership Landscape Architects 44
Victoria Ranney 52
Courtesy Harvey Rayner 211 top, 211 center
Courtesy Rios Clementi Hale Studios 113, 199
Blair Robbins and Ian Lazarus 46
Courtesy Romses Architects 123, 124, 125 all
Rooftops Canada 219 bottom right
Emily Rylander 153 left, 153 right
Jane Sebire 133 bottom
Courtesy SHED Studio, Growing Home 98, 99 top
Scott Shigley 180, 181 all
Courtesy Adam Smith/Aquapod 213
Courtesy Stantec Architects 176 top, 177 top left
Courtesy The Stop Community Food Centre 74
Tafler Rylett 172 bottom
Courtesy Teeple Architects 148, 149 all, 151
TSR Group 20
Courtesy Urban Habitat Chicago 156
Collin Varner, City Farmer 121
Mark Vetter/Jonathan Rose 174, 175 bottom
Vaughan Wascovich 50
Ben Walmsley, Parc Downsview Park, Inc. 48 right
Courtesy What if: projects Ltd. 215 top, 220 top, 220 bottom
Courtesy Windowfarms 208 top, 208 bottom
Courtesy Woolly Pocket Garden Company, Inc. 230 left
Courtesy WORK Architecture Company 90, 91 all, 92 all, 93 all, 105 bottom, 106 bottom, 107, 108 top